Electrolyte, Nutrition and Hydration Beverages for Peak Performance and Recovery

From Formulation to Field: Applications, Troubleshooting & Case Studies

A Comprehensive Resource for Formulators, Producers, Users and Consultants

About the Book

"Electrolyte, Nutrition and Hydration Beverages for Peak Performance and Recovery" is Volume 2 in *"The Smarter Nutrition Library"*, a series designed to bridge the gap between physiological science and scalable product innovation. Where Volume 1 laid the groundwork for foundational nutrient delivery, this book dives into the dynamic intersection of fluid dynamics, metabolic support, and sensory design, where hydration becomes performance architecture.

Written for formulators, product developers, and agile startups, it moves beyond generic guidelines to deliver field-tested frameworks: how to match electrolyte ratios to real-world sweat profiles, engineer osmolality for gastric comfort and cellular uptake, stabilize botanicals without compromising clarity, and layer flavor to mask bitterness before it reaches the palate, not after. Every formulation template, compatibility chart, and troubleshooting protocol reflects lessons learned in co-manufacturing facilities, pilot kitchens, and stability chambers across diverse climates and regulatory landscapes.

This is not a theory manual. It's a working document, dog-eared, annotated, and returned to, because in functional beverage development, success is measured not in peer-reviewed citations, but in shelf-stable batches, repeat purchases, and athletes who trust your drink when performance is non-negotiable.

American Industrial Technologies LLC.
Technology Transfer, Local Integration, and Industrial Development

About the Series

"The Smarter Nutrition Library" is a practical, science-led series for innovators building the next generation of functional foods and beverages. Each volume bridges deep technical expertise with real-world development, translating physiology, formulation chemistry, and sensory science into scalable, market-ready solutions. Written by formulators, for formulators, the series emphasizes actionable insight over theory, empirical validation over assumption, and human-centered design over mechanistic optimization. Volume 1 laid the foundation with dispersing additives for precision delivery; Volume 2 advances the mission by redefining hydration, not as passive repletion, but as intelligent, adaptive nourishment engineered for performance, recovery, and daily resilience.

Electrolyte, Nutrition and Hydration Beverages for Peak Performance and Recovery

From Formulation to Field: Applications, Troubleshooting & Case Studies

A Comprehensive Resource for Formulators, Producers, Users and Consultants

1st Edition

Mostafa A. Ebrahim, Ph.D.

Volume 2

American Industrial Technologies LLC.

Technology Transfer, Local Integration, and Industrial Development

Copyright

"Electrolyte, Nutrition and Hydration Beverages for Peak Performance and Recovery"
Author: Mostafa Ebrahim, PhD
Published by: American Industrial Technologies LLC.

Disclaimer

"Electrolyte, Nutrition and Hydration Beverages for Peak Performance and Recovery" is a technical reference for experienced formulators, product developers, and startup teams. The content reflects real-world development practice, not theoretical ideals, and is grounded in pilot-scale validation and commercial production experience.

It does not constitute regulatory, legal, or medical advice. Readers are solely responsible for:

- Verifying ingredient compliance (GRAS, E-number, regional approvals, heavy metal limits);
- Conducting full sensory, microbiological, and accelerated shelf-life testing;
- Adapting formulations to local labeling rules (e.g., structure/function claims, allergen declarations, %DV);
- Validating process parameters (heat, fill temperature, deaeration) under actual production conditions.

No formulation herein is a substitute for Good Manufacturing Practices (GMP), HACCP, or third-party safety review. The author and publisher expressly disclaim liability for any consequences arising from the use or misuse of this material.

Trademark and Brand Disclaimer

All trademarks, service marks, trade names, trade dress, logos, and other brand identifiers featured in this publication are the property of their respective owners. Their inclusion is solely for identification and illustrative purposes and does not imply any endorsement or affiliation with the book's author or publisher. This publication does not intend to infringe upon the rights of any trademark holder.

Intended Use Statement

Electrolyte, Nutrition and Hydration Beverages for Peak Performance and Recovery is a practical, application-focused guide tailored for beverage formulators, product developers, startup founders, and technical teams working in functional hydration, sports nutrition, and wellness beverage innovation.

It distills field-tested knowledge, not academic abstraction, into actionable frameworks for the design, stabilization, scaling, and commercialization of electrolyte and nutrition drinks. Every chapter reflects real formulation challenges, pilot-scale learnings, and market-proven strategies harnessed from commercial development cycles across global markets.

Disclosure: Acknowledging Collaborative Technology

This book has leveraged advanced writing technology as a supportive tool. While all key concepts, technical analyses, and original insights remain the author's own work, artificial intelligence assisted by suggesting relevant research, verifying facts, improving grammar, and enhancing readability through structural refinements.

Dedication

To my beloved father, whose steadfast guidance and unwavering work ethic have been a constant source of inspiration. Thank you for teaching me the true value of perseverance and the relentless pursuit of excellence.

To my caring mother, whose unconditional love and nurturing spirit have profoundly shaped me into the person I am today. Your endless encouragement has empowered me to boldly chase my dreams and passions.

To my wonderful wife, Monica, whose unwavering support and love have been my anchor. Your belief in me and our shared dreams inspire me every day.

And to my darling daughters, Fatima and Mona, whose boundless enthusiasm and insatiable curiosity remind me daily of the profound importance of wonder, discovery, and building a better future. This work is a testament to the world I hope to create for you.

This book is a humble offering to the four pillars of my life, may it serve as a small expression of my deepest, most heartfelt love and gratitude.

Acknowledgments

The knowledge in this book didn't emerge from textbooks or isolated R&D labs, it was forged on the plant floor, in pilot kitchens, and across countless failed batches that taught more than any success ever could. It grew from the hum of homogenizers at 2 a.m., the frustration of phase separation just before scale-up, and the triumph of a clear, stable, shelf-ready beverage that actually performed as promised, under real-world heat, light, and logistics.

This work is the distillation of over a decade of applied formulation, of notebooks filled with pH logs, osmolality readouts, sensory panel scores, and post-mortems on off-notes no instrument could explain. It reflects collaborations with beverage scientists, startup founders, contract manufacturers, and ingredient suppliers across North America, Europe, the GCC, and Southeast Asia, people who shared not just data, but context: how a formulation behaves when the line runs hot, when the warehouse spikes to 40 °C, or when consumers sip straight from the fridge after a 12-hour flight.

To Monica my wife, and to my beloved daughters Mona and Fatima, my partner in every sense: thank you for enduring a home office that doubled as a lab, a kitchen counter cluttered with refractometers and pH strips, and dinners delayed by "just one more stability test." Your patience, curiosity, and unwavering belief made this possible.

To my colleagues at American Industrial Technologies LLC, and to the wider community of functional beverage innovators: thank you for the debates, the blind taste tests, the shared supplier war stories, and the peer reviews that cut through jargon to demand clarity, reproducibility, and real-world relevance. You ensured this book stays rooted in what works, not theory, not trends, but validated, scalable science.

And to the next generation of formulators, product developers, and startup founders: may this guide help you move faster, not by skipping steps, but by knowing which steps truly matter. May it help you avoid the costly detours, anticipate the hidden instabilities, and build beverages that perform as brilliantly on the shelf as they do in the body.

This is not a solo effort.

It is a field manual, written by the industry, for the industry.

Forward

The functional beverage industry has entered a new era, one where hydration, nutrition, and performance are defined not only by ingredient lists but by the precision with which those ingredients are selected, balanced, and delivered. Formulating an electrolyte or nutrition beverage today requires far more than knowing which minerals to add or which flavors to blend. It requires an understanding of how chemistry, physiology, sensory science, and processing interact, sometimes subtly and sometimes dramatically, to determine whether a product succeeds on the shelf, in the body, and on the production line.

This book was written to bridge the gap between theory and practice. It distills years of hands-on experience, pilot-plant trials, stability failures, reformulations, scale-up challenges, and real-world manufacturing constraints into a clear and practical reference. Rather than offering generic guidelines or sweeping promises, it presents the boundaries of what works in electrolyte and functional hydration beverages, why certain ingredients perform beautifully at one pH but fail at another, why a seemingly minor change in osmolality can reshape taste and mouthfeel, and why processing choices often matter as much as formulation choices.

What makes this work distinctive is not only its scientific grounding but its pragmatism. The insights here reflect the realities of tight timelines, variable raw materials, cost pressures, and demanding regulatory and sensory requirements. Every recommendation has been validated through the kinds of circumstances that shape real products in real facilities, from early R&D benchwork to high-speed manufacturing.

For startups, this book offers a roadmap through a complex landscape. For seasoned developers, it provides a structured framework that captures the details often learned only through costly missteps. For anyone building the next generation of hydration and nutrition beverages, clean-label, high-performance, stable, and consumer-trusted, this guide offers a foundation strong enough to build on and flexible enough to innovate with.

Its purpose is simple: to help product developers move faster, make better decisions, and bring forward beverages that are as technically sound as they are meaningful for the consumer. If the future of hydration is smarter, cleaner, and more personalized, then this book is one of the tools that will help shape that future.

American Industrial Technologies LLC.

About the Author

About the Author

Dr. Mostafa Ebrahim is a professional technical writer and author focusing on industrial technology and knowledge transfer. He writes with extensive experience in the field of applied sciences and process optimization, and his mission is to communicate difficult-to-transfer knowledge into useful advice for entrepreneurs, artisans, and small and medium-sized companies.

American Industrial Technologies LLC.

Technology Transfer, Local Integration, and Industrial Development

Declaration of Competing Interest

I hereby declare that there are no known competing financial interests, professional affiliations, or personal relationships that could be perceived as influencing this publication, content, or recommendations presented in this work. No financial support or compensation was received from any organization that might have an interest in the outcomes of this publication.

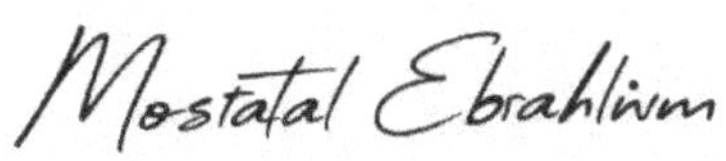

Preface

Over the past decade, functional hydration has evolved from a niche sports category into a cornerstone of wellness, performance, and daily ritual. Consumers no longer ask, "Does this hydrate?", they ask, "Does this help me think clearer, recover faster, feel calmer, and live more intentionally?" That shift has placed unprecedented pressure on formulators: to deliver not just ions and osmolality, but meaning, through taste, texture, transparency, and trust. Yet, too many technical resources remain trapped in theory: idealized equations with no regard for line fouling, elegant flavor models that ignore salt–bitterness synergy, or stability protocols that assume perfect warehouse conditions. Real-world beverage development is messier, faster, and far more human.

Electrolyte, Nutrition and Hydration Beverages for Peak Performance and Recovery is written for those who live in that reality, the formulators, product developers, startup founders, and technical managers who translate physiology into palatable prototypes, and prototypes into scalable products. It does not aim to replace peer-reviewed journals or regulatory databases; rather, it bridges them. Every chapter distills field-tested principles: why a 0.1% shift in citrate can prevent haze in a magnesium-fortified drink; how L-theanine's calm focus only emerges when paired with caffeine and the right acidity profile; why a "natural" coconut water base may require more engineering than a synthetic electrolyte system to achieve shelf stability. This is not knowledge extracted from literature alone, it is knowledge forged in co-packer trials, reformulation loops, and consumer rejection.

The structure reflects how development actually happens: we begin with physiology, because formulation must serve biology, not the other way around. We then layer in function (electrolytes, carbs, proteins, botanicals), followed by form (flavor, texture, clarity), and finally feasibility (processing, packaging, stability, regulation). Along the way, you'll find formulation templates that have run on commercial lines, compatibility charts built from real failure logs, and stability data gathered under Middle Eastern summer heat and Southeast Asian humidity, not just temperate lab conditions.

To the next generation of beverage scientists: you are not merely mixing ingredients. You are designing interfaces, between biology and behavior, between data and desire, between stress and resilience. May this book serve not as a rulebook, but as a field companion, dog-eared, annotated, and returned to often, as you build beverages that don't just perform, but resonate.

Mostafa Ebrahim
American Industrial Technologies LLC.

1. Hydration 101: The Physiology of Performance

The basis of human performance is hydration. This chapter describes the use of water to regulate temperature and transport nutrients and provide cellular functions during exercise. The readers will be informed about what occurs within the body during dehydration, the effects of osmolality in reducing performance, and how hydration strategies can be employed to keep the endurance and mind sharp. It connects physiology with applied formulation objectives by demonstrating how the design of a beverage can be initiated by knowing the pathway of the body in absorbing, distributing and retaining fluids when they become stressed.

1.1 What dehydration really does to the body

Dehydration is not just a simple sense of thirst; it is a physiological chain reaction that affects almost all body systems. In cases where the loss of fluids surpasses the intake, the plasma volume reduces, the blood becomes thicker and the heart has to work more to provide oxygen and nutrients to the active muscles. A drop in the capacity to lose heat by the body increases the core temperature, which results in fatigue, loss of coordination and slowing of cognitive response. Mild dehydration (loss in body weight as low as 1-2 percent) has the effect of compromising endurance, accuracy, and decision making which are compounded in hot weather or length of exertion. This part dissects the biological processes of dehydration, analyzing how electrolyte imbalance interferes with nerve transmission, muscle contraction and energy metabolism. To the product developers, the knowledge of these mechanisms gives the scientific basis on how to create beverages that do not only satisfy the thirsty, but also maintain physiological balance and performance in the field.

1.2 Fluid balance, osmolality, and performance drop-off

The performance invisible regulator that correlates hydration state directly with muscle efficiency and mental clarity is the fluid balance. The body is in incessant balance of water absorption and its excretion via a mechanism called osmoregulation, which is regulated by brain sensors that determine the osmolality of the plasma or the concentration of the dissolved particles in the blood and within the body fluids. Increase in osmolality, such as through losing water by sweating or other causes such as lack of sufficient fluid intake, causes water to move out of the cells to bring about a state of metabolic efficiency, which causes cellular dehydration. There is increased fatigability of muscles, decreased reaction time, and impaired thermoregulation. Even a margin of deviation of osmolality may cause a performance deterioration that is measurable, including affecting endurance, power output, and focus. The interaction of the beverage osmolality (hypotonic, isotonic, or hypertonic) with the bodily natural equilibrium, and why the best hydration beverages are developed to regulate to the rate of absorption into the body. This understanding will assist the formulators to design drinks that do not only provide hydration, but also optimum physiological performance.

1.3 How the body absorbs and retains water

1.3 Water absorption and retention in the body.

Water absorption and retention are physiological processes that require extreme finesse in the determination of the efficiency of hydration strategies. After intake, water is absorbed mostly in the small intestine whereby it enters into specialized channels known as aquaporins and sodium-glucose transporters (SGLT1) that facilitate rapid absorption of fluid. This is important because electrolytes particularly sodium, produce osmotic gradients that draw water into the bloodstream and aid in the maintenance of the extracellular fluid volume. Subsequently, the kidneys are the major regulators of water balance and alter the reabsorption rate according to the hormone cues like vasopressin (antidiuretic hormone). In case fluid intake is too low or there is a disproportion in electrolyte, the body loses water by concentrating urine, and it can temporarily keep the body hydrated at the cost of normal temperature control and metabolism.

This section describes to the beverage formulators the importance of absorption in isotonic and slightly hypotonic solutions and why the optimal combination of sodium, glucose and osmolytes promotes uptake and retention. The knowledge of these physiological principles also allows the developers to formulate hydration drinks in a manner that the body can utilize in its transport system, thereby absorbing faster, retaining more water, and providing increased performance or recovery time.

2. Electrolyte Essentials

Electrolytes are the conductors of performance that regulate the muscle contraction, nerve transmission, and fluid balance. This chapter explores how important ions, sodium, potassium, calcium, magnesium, and chloride, are in facilitating prolonged physical and intellectual functioning. It contains evidence-based information about the composition of the sweat, electrolyte loss in different climates, and the way of designing electrolyte ratios to match the physiological requirements to endurance, recovery, and daily hydration.

2.1 Role of sodium, potassium, calcium, magnesium, and chloride

Electrolytes refer to electrical circuities in the body--micronutrients which drive all impulses to the nerves and all movement in the muscles. The main ion of the extracellular fluid is sodium, which helps to balance fluids and conduct the nerve impulses. It controls thirst and aids in the intake of glucose and water in the intestines. Potassium, which is concentrated within cells, acts together with sodium to ensure the maintenance of membrane potential as well as the prevention of muscle cramps and fatigue. Calcium is a signaling molecule, which initiates the contraction of muscles, the release of hormones, and the initiation of enzymes. Magnesium is a metabolic cofactor that maintains ATP stability, muscle relaxation, energy release and heart rhythm. Although neglected, chloride neutralizes the positive ions and keeps the acid-base equilibrium and osmotic pressure.

To formulators, this section highlights the contribution made by each of the ions towards hydration efficacy and performance stability. It examines best ratios and concentrations needed to use sports, wellness, or recovery drinks- the relationship between biochemical activity and formula development. The knowledge of these interactions can be used to make electrolyte systems which can not only restore what is lost but also make physiological efficiency more efficient during exertion or in stress.

2.2 Sweat profiles and electrolyte loss in different climates

Sweat composition is as personal as a fingerprint, which is determined by genetics, nutrition, fitness status, and environment. Sweat rates and electrolyte loss may increase two times in hot rainy or high-altitude environments causing sodium and fluid loss in the body in a short time. This part divides the science of variability of sweat rate and mineral composition of perspiration, and reveals that sodium and chloride have pre-eminence, with potassium, calcium, and magnesium in less, yet physiologically significant amounts.

To the developer, the knowledge of sweat profiles enables customization of beverage formula to suit actual conditions; be it endurance sports in hot climates or light recovery hydration in cooler climates. It also presents field tested techniques to estimate the loss of electrolytes and has a fit between product assertion and physiological data to ensure effectiveness and trustworthiness in the market environment.

2.3 Balancing ions for sustained output and faster recovery

Human hydration performance is all about balance, and not mere replacement. Over representation of one of the electrolytes, particularly sodium, may cause imbalance to the fine electrical gradients that are operated by the muscles and nerves. This part is devoted to ion synergy science, the adequate ratios between sodium, potassium, calcium and magnesium maintain muscle performance, prevent fatigue and enhance the recovery process after physical activity.

Formulation wise, it offers useful information on how to be able to engineer the elelectrolyte matrices that resemble the plasma composition of the body, so that the best absorption and retention is realized. It also points out the impact of ionic balance on osmolality, perception of taste and fluid kinetics. Formulators have the power to create the beverage that helps the body stay in a rhythm hydrated and stable and ready to perform optimally by mastering the art of electrolyte harmony.

3. Types of Hydration Beverages

Hydration beverages are broadly categorized by their tonicity, the concentration of dissolved solutes (mainly electrolytes and carbohydrates) relative to that of body fluids. This property determines how quickly water and nutrients move across intestinal membranes, directly influencing absorption rate, energy delivery, and performance outcome. Understanding the physiological behavior of isotonic, hypotonic, and hypertonic formulations allows formulators to strategically design beverages that meet specific functional goals, whether for rapid hydration, sustained endurance, or post-exercise recovery.

3.1 Isotonic Beverages (Direct Rivals)

Isotonic drinks dominate the performance hydration category, making them the closest and most competitive rivals for any new electrolyte or functional hydration product. Their balance of carbohydrates, electrolytes, and acidity is engineered to match the body's fluid absorption rate , a claim that has shaped consumer expectations for decades. This section examines the core attributes, strengths, and limitations of isotonic beverages, helping formulators understand exactly what they must match, exceed, or differentiate from. By analyzing flavor profiles, osmolality targets, ingredient norms, and market positioning strategies, you'll gain a clear view of the benchmark standards that define today's direct competitors in the isotonic space.

3.1.1 Definition

Isotonic beverages have an osmolality similar to human blood plasma, typically around 270–330 mOsm/kg. This balance allows for optimal fluid absorption while simultaneously providing a steady supply of carbohydrates and electrolytes.

3.1.2 Mode of Action

Because their solute concentration mirrors that of body fluids, isotonic drinks enable rapid rehydration without causing gastric distress or delaying gastric emptying. They deliver a moderate carbohydrate load (usually 6–8%) for quick energy replenishment, making them ideal during prolonged sports or heavy exertion.

3.1.3 Key Benefits

- Balanced hydration and energy delivery in one system.
- Rapid absorption with minimal gastric discomfort.
- Maintains electrolyte equilibrium under moderate-to-high exertion.
- Familiar taste and mouthfeel, strong consumer acceptance.
- Flexible formulation: compatible with flavor systems, vitamins, and mild protein inclusion.

3.1.4 Drawbacks / Limitations

- Moderate sugar level (6–8%) can be too high for low-calorie positioning.
- May not hydrate as fast as hypotonic solutions.
- Overuse during light activity can add unnecessary calories or sodium.

3.1.5 Ideal Applications

- Competitive sports and endurance training.
- Daily performance hydration.
- Mass-market "sports drink" category (Gatorade, Powerade, Prime).

3.1.6 Commercial Examples

Classic isotonic brands include Gatorade, Powerade, and Lucozade Sport, which dominate the mainstream performance market. Newer entries, such as BodyArmor and Prime Hydration, focus on clean-label ingredients and natural electrolyte sources while maintaining isotonic balance for broad consumer appeal.

3.2 Hypotonic Beverages - Functional Segment (Rapid Absorption)

Hypotonic beverages occupy a unique space in the hydration landscape, positioned for moments when speed matters more than energy density. Engineered with lower solute concentrations, these drinks move across the gastric barrier quickly, delivering fluid replenishment with minimal digestive load , a property increasingly valued in heat stress environments, endurance sports, and wellness-focused daily hydration. This section explores how the formulation architecture of hypotonic systems shapes their functional advantage, and how brands can position them as clean, efficient, and science-backed alternatives to traditional sports drinks. By understanding the nuances of rapid absorption and consumer expectations, formulators can craft hypotonic beverages that stand out for their clarity of purpose and performance.

3.2.1 Definition

Hypotonic beverages contain lower solute concentrations than blood plasma, generally below 270 mOsm/kg. Their reduced osmolality enables water to move quickly from the digestive tract into the bloodstream.

3.2.2 Mode of Action

Designed for rapid rehydration, hypotonic drinks are ideal when fluid replacement is the primary goal rather than energy delivery. Their lower sugar and electrolyte content allow for faster gastric emptying and quicker absorption, making them suitable for hot environments, short-duration training, or post-activity recovery.

3.2.3 Key Benefits

- Fastest fluid absorption rate; ideal for rehydration.
- Light, refreshing taste with low sugar and sodium.
- Gentle on the stomach, minimal gastric load.
- Perfect for warm climates or short-duration exertion.
- Supports “clean-label” and “low-calorie” market trends.

3.2.4 Drawbacks / Limitations

- Lower carbohydrate and energy content, not suitable for endurance fueling alone.
- Less effective for glycogen replenishment.

- Requires careful balancing to maintain palatability and stability.
- Electrolyte concentrations may be too low for heavy sweat loss.

3.2.5 Ideal Applications

- Fitness, yoga, wellness hydration.
- Heat-stress recovery and post-workout rehydration.
- Lifestyle "functional water" products (Nuun, Hydralyte, Propel).

3.2.6 Commercial Examples

Products such as Hydralyte, SOS Hydration, and Nuun Sport (low-carb tablets) represent this segment, focusing on quick fluid uptake, clean-label positioning, and low-calorie formulations. They appeal to wellness-oriented consumers seeking functional hydration without excess sugar or heaviness.

3.3 Hypertonic Beverages (Recovery Segment)

Hypertonic beverages serve a distinct role in the performance spectrum, delivering concentrated carbohydrates, electrolytes, and functional nutrients to accelerate post-exercise recovery. With higher solute loads designed to support glycogen replenishment and sustained nutrient delivery, these systems address the physiological demands that follow intense training or prolonged exertion. This section unpacks the science that gives hypertonic drinks their recovery edge, and examines how brands can position them as purposeful, nutrient-dense solutions rather than simply "high-sugar" formulations. By connecting formulation design with real recovery needs, this chapter helps define how hypertonic beverages can stand apart as targeted tools for rebuilding, restoring, and preparing the body for what's next.

3.3.1 Definition

Hypertonic beverages have a higher solute concentration than blood plasma, often exceeding 330 mOsm/kg. They contain elevated carbohydrate levels, typically 8–15%, and are designed primarily for energy replenishment rather than hydration.

3.3.2 Mode of Action

Due to their high solute load, hypertonic beverages are absorbed more slowly and can temporarily draw water into the intestinal lumen, delaying hydration but providing dense energy for glycogen restoration. They are best consumed after intense activity to accelerate carbohydrate recovery or as part of endurance fueling systems when used alongside water or hypotonic drinks.

3.3.3 Key Benefits

- Dense carbohydrate content, excellent for energy replenishment.
- Effective post-exercise glycogen restoration.
- Can deliver high nutrient loads (proteins, amino acids, recovery additives).
- Useful as part of multi-stage hydration/energy protocols.

3.3.4 Drawbacks / Limitations

- Slower absorption; may delay hydration.
- Can cause gastrointestinal discomfort or bloating if consumed during activity.
- May draw water into the gut, increasing dehydration risk during exertion.
- Typically, high in sugar, limits "health" positioning.

3.3.5 Ideal Applications

- Post-workout recovery drinks.
- Endurance fueling systems (marathons, triathlons).
- Meal-replacement or recovery beverage categories (SIS Rego, Perpetuem).

3.3.6 Commercial Examples

Leading examples include Hammer Nutrition Perpetuem, SIS Rego Rapid Recovery, and concentrated energy gels or carbohydrate drinks used in endurance sports. These products are typically positioned as part of a multi-stage performance strategy, hydration during exercise via isotonic or hypotonic drinks, followed by hypertonic formulations for post-activity energy restoration.

3.4 Key Insights for Product Developers

In a category defined by rapid innovation and evolving consumer expectations, product developers need more than technical proficiency, they need strategic insight. This section distills the essential lessons drawn from the competitive hydration landscape, highlighting the formulation, sensory, and market cues that separate successful beverages from those that fail to resonate. From aligning osmolality with use-case to balancing flavor, function, and label positioning, these insights provide a practical framework for making informed decisions throughout development. Here, formulation strategy meets market reality, giving developers a clear roadmap for creating hydration products that stand out with purpose, performance, and precision.

- Hypotonic drinks dominate the wellness and clean hydration segment.
- Isotonic drinks remain the mainstream sports performance standard.
- Hypertonic drinks target recovery and endurance use, often paired with hypotonic options in staged hydration systems.
- The optimal formulation strategy often blends features, for example, isotonic systems for in-activity hydration and hypotonic variants for post-activity fluid restoration.

4. Market Positioning

In a crowded and fast-moving beverage landscape, even the most technically sound formula needs a clear, competitive identity to succeed. Market positioning is where science meets perception, where functional performance, consumer expectations, and brand storytelling converge to define how a hydration or nutrition drink is understood, chosen, and valued. This chapter explores how to translate formulation choices into compelling market narratives, how to differentiate within saturated categories, and how to align product benefits with the needs of athletes, lifestyle consumers, wellness seekers, and emerging functional-beverage segments. Here, you'll learn to position a beverage not just as a drink, but as a purposeful solution that earns its place on the shelf and in the consumer's daily routine.

4.1 Target Market Positioning by Beverage Type

Different beverage types speak to different needs, occasions, and consumer mindsets, and effective positioning begins with understanding those distinctions. Whether the product is a hypotonic rehydration drink, an isotonic performance formula, a recovery system, or a wellness-focused functional water, each category carries its own expectations for taste, nutrition, claims, and experience. This section outlines how to align a beverage's composition with the audience it serves, translating technical formulation choices into clear consumer value. By mapping beverage type to target user, use-case, and competitive landscape, you'll learn to position each product with precision, relevance, and market clarity.

4.1.1 Isotonic Beverages, The Mainstream Performance Segment

Isotonic beverages dominate the sports and active lifestyle market, serving both athletes and everyday fitness consumers. They are designed for individuals who want balanced hydration with a moderate energy boost, ideal for team sports, gym workouts, and endurance activities. Typical consumers are young adults and recreational athletes seeking a recognizable, flavorful product that performs under pressure.

From a branding and marketing perspective, isotonic drinks are positioned as "Performance Hydration with Energy Support." They promise fluid and electrolyte replacement with a touch of carbohydrate-driven energy, making them versatile and familiar. Brands like Gatorade, Powerade, and Prime Hydration define this segment through vibrant packaging, mass-market distribution, and trust built over decades of visibility in sports culture.

Isotonic beverages generally sit in the mid-range price tier (USD $1–2 per unit), with strong consumer loyalty and broad appeal. They are ideal for line extensions, such as low-calorie, natural electrolyte, or caffeine-enhanced variants. The key commercial advantage lies in scalability, these products fit easily into mainstream retail channels while leaving room for innovation through cleaner, more transparent formulations.

4.1.2 Hypotonic Beverages, The Functional and Wellness Hydration Category

Hypotonic beverages target the wellness and recovery segment, appealing to consumers who prioritize rapid absorption, low calorie content, and clean-label ingredients. The target demographic includes health-conscious adults, light exercisers, travelers, and even medical users requiring rehydration. Their appeal extends beyond sports, reaching office workers and seniors who seek everyday hydration solutions that go beyond plain water.

Their brand identity centers on "Rapid Hydration and Electrolyte Balance." Hypotonic drinks are typically lighter in taste, gentler on digestion, and lower in sugar than their isotonic counterparts. Products like Hydralyte, Nuun Sport, and Propel embody this category, often marketed as tablets, electrolyte powders, or bottled "functional waters."

Positioned in the mid to premium pricing range (USD $1.50–3 per serving), these beverages thrive in health-oriented retail spaces, e-commerce, and travel outlets. They benefit from associations with natural minerals, vitamin fortification, and plant-based electrolyte sources. For product developers, the opportunity lies in merging science-backed hydration performance with clean-label lifestyle branding, a fast-growing niche driven by modern wellness consumers.

4.1.3 Hypertonic Beverages, The Recovery and Endurance Nutrition Niche

Hypertonic beverages occupy a specialized space in the performance nutrition market, catering to endurance athletes, marathoners, cyclists, and competitive sports professionals. Their high carbohydrate concentration makes them ideal for post-exercise recovery and glycogen replenishment, rather than rapid hydration. The typical consumer is performance-focused, data-driven, and willing to invest in advanced, science-based formulations.

Their market positioning emphasizes "Energy and Recovery Fuel." Hypertonic drinks are often sold as part of a structured performance system, paired with hypotonic or isotonic products for before, during, and after exercise. They deliver dense nutrition and measurable physiological benefit, marketed through evidence-based claims and technical branding.

Falling into the premium to specialized price tier (USD \$2–5 per unit or \$25–40 per multipack), these products appeal to a smaller but loyal market segment with high repeat-purchase potential. Notable examples include SIS Rego Rapid Recovery, Hammer Nutrition Perpetuem, and advanced carbohydrate-protein recovery formulas.

For formulators and startups, this segment offers strong differentiation through innovative ingredient systems, such as cluster dextrin, BCAAs, and adaptogens, that elevate perceived performance and justify higher margins. While niche, the hypertonic category defines the cutting edge of functional beverage innovation, blending endurance nutrition with hydration science.

Table 4.1.3: Summary Comparison

Type	Core Market Segment	Primary Consumer Profile	Brand / Marketing Positioning	Price Tier	Strategic Opportunities
Isotonic Beverages	**Mainstream Sports**	• Athletes and fitness enthusiasts	• Marketed as "**Performa**	**Mid-range mass market**	• Appeals to both active consumers

	& Active Lifestyle	• Recreational gym users • Team sports participants • Young adults seeking energy and hydration	**nce Hydration with Energy Support**." • Balances electrolyte replenishment and light fueling. • Positioned as the "everyday sports drink." • Familiar flavors, bold branding, and broad retail distribution (convenience and supermarkets).	(USD $1–2 per unit)	and casual users. • Strong potential for clean-label reformulation (natural electrolytes, low sugar). • Line extensions: low-calorie or caffeine-enhanced variants.
Hypotonic Beverages	**Wellness, Recovery, and Functional Hydration**	• Health-conscious consumers • Light exercisers, yoga, and pilates	• Marketed as "**Rapid Hydration and Electrolyte Balance**." • Focuses	**Mid to premium** (USD $1.50–3 per unit; higher per serving for	• Ideal for functional beverage and health supplement crossover markets.

		users • Office workers and travelers • Medical or senior hydration users	on fast absorption, light taste, and low calorie. • Often labeled as “hydration tablets,” “electrolyte water,” or “wellness hydration.” • Appeals to clean-label and low-sugar lifestyles.	powders/tablets)	• Compatible with natural, plant-based, or vitamin-fortified positioning. • Opportunities in e-commerce and travel retail channels.
Hypertonic Beverages	**Endurance, Recovery, and Specialized Performance Nutrition**	• Endurance athletes (runners, cyclists, triathletes) • Sports professionals and teams • Post-training recovery users • Serious	• Marketed as “**Energy and Recovery Fuel**.” • Emphasizes carbohydrate density and glycogen replenishment. • Often sold as part of	**Premium to specialized** (USD $2–5 per unit or $25–40 per multipack)	• High potential for nutrition–performance fusion products (protein + carbs + electrolytes). • Market differentiation through advanced ingredient

	fitness consumers	multi-phase systems (e.g., pre-, during-, post-workout). • Positioned as a technical performance product, not a casual beverage.		systems (e.g., cluster dextrin, BCAAs, adaptogens). • Niche but loyal consumer base, high repeat purchase potential.

Summary Insight

- Isotonic beverages own the mass-market performance space, easily scalable, broad consumer appeal, and recognizable branding.
- Hypotonic beverages are leading growth in the functional hydration and clean-label wellness markets, especially for low-calorie and natural ingredient claims.
- Hypertonic beverages dominate the performance nutrition and recovery niche, often tied to premium pricing and specialized athletic communities.

Together, these three categories create a multi-tiered market strategy:
→ Hypotonic for daily wellness,
→ Isotonic for performance hydration, and
→ Hypertonic for energy and recovery.

4.2 Integrated Market Strategy - Positioning Across the Hydration Spectrum

In today's functional beverage market, success rarely comes from a single formula, it comes from building a tiered product portfolio that meets diverse hydration and performance needs. Isotonic, hypotonic, and hypertonic beverages each serve a distinct physiological and commercial purpose, and together they form a complete hydration ecosystem that supports consumers before, during, and after activity.

Isotonic beverages act as the backbone of this system, accessible, familiar, and designed for active hydration during exercise or daily performance. Hypotonic beverages extend the brand into the wellness space, appealing to health-conscious consumers and new demographics who value clean-label, low-calorie functionality. Meanwhile, hypertonic beverages elevate the brand into the premium and professional performance segment, offering energy-dense recovery and endurance solutions for specialized users.

For formulators and startups, the strategic advantage lies in product layering: creating a cohesive brand family that spans hydration, energy, and recovery while maintaining consistent sensory identity and visual design. This approach not only broadens market reach but also builds consumer trust and lifecycle engagement, allowing the same brand to accompany the user from everyday wellness to elite performance.

In essence, these three beverage types aren't competitors; they're complementary pillars of a unified hydration strategy, one that integrates physiological science with modern consumer expectations for personalization, performance, and clean nutrition.

5. Fuel Meets Fluid: Nutritional Synergy

This chapter explores how hydration and nutrition intersect to drive metabolic performance. Beyond water and electrolytes, it details how carbohydrates, proteins, and amino acids work together to enhance fluid uptake, stabilize blood sugar, and accelerate muscle repair. Readers will gain a deeper understanding of "metabolic hydration," where nutrient design amplifies fluid efficiency and energy delivery for both athletic and functional beverage systems.

5.1 How carbohydrates enhance hydration and energy availability

There are two roles of carbohydrates in hydration drinks, carbohydrates are not only the energy sources but also the hydration promoters. Carbohydrates are effective potent fluid absorbents in the small intestine when formulated properly, using the sodium-glucose co-transport (SGLT1) active method which actively draws water into bloodstream in association with glucose and sodium. Such synergy does not only speed up the process of rehydration but also restores the glycogen reserves used up during exercise. The carbohydrate type and content of a beverage spell out the difference between sudden energy and long-lasting power, simple sugars such as glucose and sucrose are fast and easy sources of energy, whereas maltadextrin and cluster dextrin are stable sources of energy with minimum gastrointestinal strain.

5.2 Protein and amino acid roles in post-exercise recovery

Proteins and amino acids play a central role in the post-exercise recovery process because they supply the raw materials needed for repairing muscle fibers that experience micro-damage during physical activity. Exercise, especially resistance and high-intensity training, triggers a cascade of adaptive responses that increase protein turnover and stimulate muscle protein synthesis. Providing readily available amino acids during this window helps shift the body from a catabolic to an anabolic state, accelerating tissue repair and supporting long-term performance adaptations.

Amino acids also influence metabolic recovery by assisting in glycogen replenishment and supporting immune function, both of which may be compromised after strenuous training. Branched-chain amino acids (BCAAs) are particularly well-studied for their role in reducing muscle protein breakdown and modulating central fatigue pathways. Meanwhile, conditionally essential amino acids like glutamine and arginine contribute to immune resilience, nitrogen balance, and cellular repair. Together, these functions make amino acids essential tools for restoring readiness after demanding exercise.

In beverage formulations, proteins and free amino acids contribute to more than just nutritional recovery, they also interact with acidity, minerals, and flavor systems in ways that influence stability and sensory outcomes. The choice of protein type (whey, collagen, plant-based isolates) or amino acid blend affects solubility, viscosity, and flavor masking requirements. A well-designed recovery beverage therefore integrates both functional nutrition and formulation science: delivering the right amino acid profile at the right concentration, in a system optimized for clarity, taste, and consistent performance across its shelf life.

5.3 The science of "metabolic hydration"

Metabolic hydration refers to the integration of fluid intake with the metabolic processes that govern energy production, electrolyte balance, and cellular recovery. While traditional hydration focuses primarily on replacing water and salts lost through sweat, metabolic hydration looks deeper, examining how nutrients, cofactors, and metabolic substrates influence the body's ability to retain fluids, clear metabolic byproducts, and restore physiological equilibrium after exertion. This perspective recognizes hydration as an active biochemical process rather than a passive replenishment of fluids.

Central to metabolic hydration is the interplay between electrolytes, carbohydrates, amino acids, and organic acids that drive cellular transport mechanisms. Sodium and glucose, for example, work in tandem through the sodium–glucose cotransporter (SGLT1) to enhance water absorption in the small intestine.

Concurrently, metabolic intermediates such as citrate and malate participate in buffering systems that reduce acidosis and facilitate energy pathway recovery. Amino acids, including alanine and glutamine, contribute by supporting osmotic balance and fueling gluconeogenesis, further tightening the connection between hydration and metabolic regulation.

In beverage design, the principles of metabolic hydration guide formulators toward nutrient systems that do more than restore fluid, they help optimize the biochemical environment for performance recovery. The selection and ratio of electrolytes, sugars, amino acids, and buffering agents determine how effectively a beverage supports mitochondrial function, acid–base stability, and post-exercise nutrient transport. When engineered correctly, metabolic hydration systems create a physiological advantage: faster rehydration, smoother metabolic recovery, and improved readiness for subsequent physical activity.

6. Performance Under Pressure

Real-world conditions, heat, humidity, altitude, and intensity, affect hydration demand and performance outcomes. This chapter shows how physiological stress alters fluid loss and electrolyte needs. Through evidence-based discussion, it provides practical hydration strategies for endurance, power, and precision sports, while highlighting the importance of personalized hydration planning for varying body types, climates, and activity durations.

6.1 Thermoregulation and heat stress

Thermoregulation is the body's ability to maintain a stable internal temperature despite changes in environmental conditions and metabolic heat production. During exercise, especially in warm or humid environments, the rate of heat generation can exceed the body's capacity to dissipate it. Sweating becomes the primary cooling mechanism, but this process also leads to substantial fluid and electrolyte loss. When fluid losses surpass the body's ability to maintain blood volume and evaporative cooling, thermoregulatory efficiency declines, increasing the risk of heat-related fatigue and injury.

Heat stress places additional strain on cardiovascular and metabolic systems. As dehydration sets in, plasma volume decreases, forcing the heart to work harder to circulate blood and cool the body. Core temperature rises more rapidly, and perceived exertion increases even at lower workloads. These physiological shifts impair muscle function, reduce cognitive performance, and increase susceptibility to cramping, heat exhaustion, or heat stroke. The interplay between hydration status and heat tolerance highlights why fluid replacement is essential, not just for performance, but for safety.

Effective hydration strategies aim to preserve plasma volume, electrolyte balance, and sweating capacity. Sodium, in particular, plays a critical role in maintaining fluid distribution and preventing hyponatremia during prolonged sweating. Beverages designed for heat-stress environments often prioritize moderate carbohydrate levels, elevated sodium concentrations, and rapid absorption to match sweat rates. By supporting thermoregulation through thoughtful formulation, hydration systems can help athletes maintain cooling efficiency and sustain performance in high-temperature conditions.

6.2 Hydration strategies for endurance, power, and precision sports

Different athletic disciplines impose unique physiological demands, requiring hydration strategies that align with the sport's intensity, duration, and performance objectives. Endurance sports, such as running, cycling, and triathlon, create sustained fluid and electrolyte losses through prolonged sweating. Athletes in these disciplines benefit from steady intake of water, sodium, and carbohydrates to maintain plasma volume, fuel availability, and thermoregulation. Isotonic or slightly hypotonic beverages are typically favored for their balance of rapid absorption and sustained energy delivery.

Power and strength sports present a contrasting hydration profile. Although total sweat losses may be lower, short bursts of high-intensity effort increase metabolic heat production and place sudden demands on neuromuscular function. Electrolytes such as sodium, potassium, and magnesium support muscle firing, while moderate carbohydrate intake can enhance explosive performance and delay fatigue. Hydration for power athletes focuses on maintaining neuromuscular readiness, minimizing cramping risk, and preventing subtle dehydration that can impair force output.

Precision and skill-based sports, such as tennis, golf, climbing, or motorsports, require sharp cognitive function, fine motor control, and decision-making accuracy. Even mild dehydration can impair reaction time, concentration, and visual processing. For these athletes, hydration strategies incorporate smaller, more frequent fluid intakes with balanced electrolytes and low to moderate carbohydrates to stabilize blood glucose and support mental performance. Flavor profile, ease of consumption, and gastrointestinal comfort also play a major role in adherence during competition.

6.3 Altitude, humidity, and fatigue factors

Environmental conditions such as altitude and humidity significantly influence hydration requirements and fatigue onset. At altitude, reduced oxygen availability increases ventilation rate, leading to greater respiratory fluid loss. The dry, thin air accelerates dehydration even in cooler temperatures, while altitude-induced diuresis further reduces total body water. These combined effects can impair aerobic capacity, slow recovery, and exacerbate the early symptoms of altitude sickness. Hydration strategies at altitude emphasize increased fluid intake, elevated electrolyte support, and the use of carbohydrate systems that spare oxygen during metabolism.

Humidity creates a different challenge by reducing the efficiency of evaporative cooling. When sweat cannot evaporate effectively, the body accumulates heat more quickly, driving earlier fatigue and greater cardiovascular strain. Athletes may experience heavy sweat losses without effective cooling, increasing the risk of dehydration, heat exhaustion, and performance decline. Hydration solutions for humid environments often include higher sodium concentrations, slightly lower carbohydrate loads, and rapid-absorption systems that support faster fluid turnover.

Fatigue itself can be both a cause and consequence of inadequate hydration. Dehydration disrupts neuromuscular coordination, slows cognitive processing, and increases perceived effort, all of which amplify fatigue during training or competition. Environmental stressors compound these effects by challenging thermoregulation and energy metabolism. Understanding how altitude and humidity shape hydration needs allows athletes and formulators to build targeted strategies that minimize fatigue, preserve performance, and maintain physiological balance across diverse conditions.

7. Formulation Fundamentals

Every successful beverage starts with a clear purpose and formulation blueprint. This chapter walks readers through the process of turning an idea into a functional prototype, from defining the product goal (rehydration, performance, or recovery) to balancing taste, efficacy, and stability. It introduces the key parameters, pH, osmolality, sweetness, and mouthfeel, that form the backbone of every high-performing hydration formula.

7.1 From concept to functional prototype

Transforming a beverage idea into a functional product begins with defining intent and translating it into formulation logic. Every great hydration drink starts with a problem to solve, whether that's faster fluid absorption, cleaner energy delivery, or better recovery after intense effort. The concept phase involves identifying the target user group, understanding their physiological needs, and aligning the product's purpose with both scientific function and market demand. For example, a hydration drink for athletes competing in humid climates will differ substantially from one designed for daily wellness or clinical rehydration.

Once the concept is defined, the formulation blueprint takes shape. This includes selecting the base water system, determining the electrolyte mix, and integrating carbohydrates, amino acids, or active nutrients as needed. Developers must evaluate parameters such as pH, osmolality, and total dissolved solids, as these govern absorption efficiency and product stability. The prototype stage transforms the formulation plan into a tangible test batch, where the theoretical balance between hydration performance, sensory appeal, and cost is first tested.

From there, iteration and optimization refine the prototype, adjusting sweetness, mouthfeel, and clarity while ensuring that function remains uncompromised. Early stability testing and sensory trials help identify ingredient interactions and flavor persistence under storage conditions. By following a structured path from idea to validated prototype, formulators can efficiently bridge the gap between innovation and market readiness, ensuring that every beverage leaving the lab is both physiologically effective and commercially viable.

7.2 Defining the purpose: rehydration, performance, or recovery

Every hydration beverage must be designed around a clear physiological and functional purpose, the foundation on which all formulation and marketing decisions are built. Without a defined purpose, even the most advanced ingredient system risks falling short of its intended effect. In hydration science, three distinct formulation objectives dominate: rehydration, performance, and recovery. Each serves a unique physiological demand and requires a different balance of electrolytes, carbohydrates, and active components.

Rehydration beverages focus on rapid fluid and electrolyte replenishment. These are typically hypotonic systems, optimized for fast absorption through low osmolality and moderate sodium levels. They restore plasma volume efficiently after fluid loss from exercise, heat exposure, or illness. The priority here is speed of hydration and electrolyte precision rather than calorie contribution.

Performance beverages, on the other hand, are engineered to sustain energy and endurance during ongoing activity. These drinks usually adopt an isotonic formulation, balancing fluid uptake with a moderate carbohydrate load (4–8%) and electrolytes that mirror sweat composition. Their role is to maintain blood glucose levels, delay fatigue, and support thermoregulation during prolonged exertion.

Finally, recovery beverages shift focus from immediate hydration to restoration and repair. Typically, hypertonic or multi-phase in nature, they contain higher carbohydrate concentrations, amino acids, peptides, and sometimes protein to accelerate glycogen replenishment and muscle recovery. Sodium and potassium are included to support post-exercise rehydration, while magnesium and calcium aid neuromuscular recovery.

By clearly defining whether a beverage aims to rehydrate, sustain, or restore, formulators can strategically select ingredient types, concentrations, and tonicity profiles that match the body's needs at each performance stage. This clarity of purpose ensures that every product not only meets a physiological function but also occupies a distinct and credible position in the competitive hydration market.

7.3 Balancing taste, efficacy, and stability

Creating a successful hydration beverage is as much an art as it is a science. The most technically sound formulation can fail commercially if it doesn't deliver a pleasant sensory experience, while a great-tasting drink is worthless if it can't perform physiologically or remain stable over time. The true skill of beverage formulation lies in achieving the three-way balance between taste, efficacy, and stability, a harmony that determines both product performance and consumer acceptance.

Taste drives repeat purchase. The flavor profile must complement the function, not mask it. Sodium and other electrolyte salts can introduce salinity or bitterness, which must be balanced with acidity, sweetness, and aroma. Selecting the right sweetener system, whether natural (stevia, monk fruit, sugar) or synthetic (sucralose, acesulfame-K), requires careful calibration to achieve refreshment without cloying sweetness or aftertaste. Acids such as citric or malic acid enhance perceived freshness and contribute to microbial control, while natural flavors and colorants help convey brand identity and psychological cues of hydration and energy.

Efficacy ensures that the drink performs as intended. This depends on maintaining the correct osmolality, electrolyte ratios, and nutrient bioavailability. Over-flavoring or excessive carbohydrate addition can compromise absorption efficiency, while under-dosing functional ingredients risks making the beverage physiologically irrelevant. The goal is to deliver measurable hydration and performance benefits that align with the product's claimed purpose, whether that's rehydration, endurance fueling, or recovery enhancement.

Stability safeguards both taste and function across the product's shelf life. Temperature fluctuations, light exposure, and ingredient incompatibilities can degrade vitamins, oxidize flavors, or cause precipitation. Choosing compatible emulsifiers, pH stabilizers, and packaging materials is essential. Developers often use accelerated shelf-life testing to evaluate flavor retention, color consistency, and electrolyte stability over time.

Ultimately, the most successful hydration beverages achieve scientific performance with consumer enjoyment. They taste crisp and refreshing, perform reliably under physiological stress, and remain visually and chemically stable from bottling to consumption. This balance of taste, efficacy, and stability is what turns a lab-tested formulation into a trusted product, one that performs in both the market and thc body.

Practical formulation design transforms hydration science into real-world, manufacturable products. This section presents illustrative formulation frameworks that demonstrate how electrolyte balance, carbohydrate load, and tonicity can be adapted to achieve distinct physiological outcomes. Each example highlights key formulation logic, ingredient function, concentration, and processing compatibility, that can serve as a reference point for developers when creating or optimizing their own beverage systems.

8. Formulation Examples

Practical formulation design transforms hydration science into real-world, manufacturable products. This section presents illustrative formulation frameworks that demonstrate how electrolyte balance, carbohydrate load, and tonicity can be adapted to achieve distinct physiological outcomes. Each example highlights key formulation logic, ingredient function, concentration, and processing compatibility, that can serve as a reference point for developers when creating or optimizing their own beverage systems.

8.1 Rehydration Beverage (Hypotonic System)

Rehydration beverages are designed for fast and efficient fluid absorption, making them the cornerstone of hydration science. As hypotonic systems, they contain a lower solute concentration than blood plasma, allowing water and electrolytes to pass rapidly through the intestinal wall into the bloodstream. This category focuses on restoring the body's fluid balance after dehydration caused by heat, exertion, or illness. In this section, readers will explore how to optimize electrolyte ratios, choose appropriate carbohydrate levels for improved absorption, and maintain light, refreshing sensory profiles suitable for quick consumption. Hypotonic drinks serve not only athletes and field workers but also wellness and medical users, bridging the gap between clinical hydration and functional daily wellness.

8.1.1 Basic Electrolyte Rehydration Drink

The basic electrolyte rehydration drink represents the simplest yet most essential formulation within hypotonic systems. It delivers the core electrolytes, sodium, potassium, and magnesium, in precise proportions to quickly restore hydration and prevent cramping or dizziness. This formulation minimizes carbohydrate load to maintain low osmolality while using natural acidity and mild sweetness to enhance palatability. Ideal for mild dehydration recovery after exercise or outdoor work, it sets the foundation for more advanced formulations by balancing speed of absorption, electrolyte accuracy, and drinkability. Readers will learn how to fine-tune salt ratios and flavor adjustments to create fast-acting hydration solutions that remain consumer-friendly and physiologically effective.

Table 8.1.1: Basic Electrolyte Rehydration Drink

Component	Typical Level (per 1 L)	Function / Rationale
Purified water	q.s. to 1000 mL	Base fluid
Sodium chloride	0.45 g (~180 mg Na^+)	Replaces sweat sodium
Potassium citrate	0.2 g (~80 mg K^+)	Aids muscle function
Glucose	2% w/v	Improves intestinal absorption
Citric acid	0.1% w/v	Balances pH, adds tartness
Natural lemon flavor	0.1% w/v	Enhances palatability
Osmolality	200–230 mOsm/kg	Rapid rehydration with minimal GI load

8.1.2 Oral Rehydration Formula (WHO-Type)

The WHO-type oral rehydration formula is the gold standard of clinical hydration, developed for medical and humanitarian applications where dehydration poses critical risks. This scientifically validated formula demonstrates how specific sodium, glucose, and citrate ratios can restore electrolyte and fluid balance with exceptional efficiency. The section discusses the principles behind sodium-glucose co-transport, which drives water absorption even during severe dehydration, and how this mechanism has been adapted from healthcare use to performance and wellness beverages. By studying this model, formulators gain a deeper understanding of how to design data-driven hydration systems that combine clinical efficacy with consumer market appeal.

Table 8.1.2: Oral Rehydration Formula (WHO-Type)

Component	Typical Level (per 1 L)	Function / Rationale
Water	q.s. to 1000 mL	Base
Sodium chloride	2.6 g	Sodium replacement
Trisodium citrate	2.9 g	Buffers acidity, aids absorption
Potassium chloride	1.5 g	Replenishes K^+ losses
Glucose (anhydrous)	13.5 g	Facilitates sodium–glucose co-transport
Osmolality	245 mOsm/kg	Clinically proven oral rehydration blend

8.1.3 Natural Mineral-Based Hydration Water

Natural mineral-based hydration waters represent a clean-label evolution of rehydration drinks. Instead of relying solely on synthetic salts, they harness naturally occurring electrolytes from mineral water, sea salt, or coconut water to deliver gentle, refreshing hydration with a premium wellness appeal. This section explores how natural mineral profiles can provide functional hydration while appealing to health-conscious consumers seeking purity and simplicity. It also addresses formulation challenges such as mineral variability, flavor consistency, and labeling claims. Readers will discover how to balance natural authenticity with technical performance, positioning these products within the growing "functional water" category that blends hydration efficacy with lifestyle branding.

Table 8.1.3: Natural Mineral-Based Hydration Water

Component	Typical Level (per 1 L)	Function / Rationale
Spring/mineral water	q.s.	Natural electrolyte source
Sea salt extract	0.4 g	Sodium + trace minerals
Coconut water concentrate	3% v/v	Natural K^+ source and sweetness
Citric acid	0.1%	pH and flavor control
Fruit essence (lime/cucumber)	As needed	Natural flavor profile
Osmolality	220–250 mOsm/kg	Clean-label "hydration water" format

8.1.4 Vitamin-Enhanced Hydration Drink

The vitamin-enhanced hydration drink elevates the hypotonic format by integrating micronutrients that support energy metabolism, immune function, and recovery. These beverages combine rapid electrolyte absorption with essential vitamins, particularly C and B-complex, to deliver both physiological hydration and added nutritional value. This section explains how to incorporate heat- and light-sensitive vitamins without compromising stability or clarity, and how to maintain low osmolality while achieving perceptible functional benefits. The resulting drinks appeal to modern consumers who want more than hydration, they want hydration with purpose, offering freshness, vitality, and a clear sense of performance enhancement in every sip.

Table 8.1.4: Vitamin-Enhanced Hydration Drink

Component	Typical Level (per 1 L)	Function / Rationale
Purified water	q.s.	Base
Sodium chloride	0.6 g	Fluid retention
Potassium citrate	0.25 g	Muscle and nerve function
Glucose + fructose	2.5%	Absorption enhancement
Vitamin C	100 mg	Antioxidant and freshness cue
B-complex vitamins	Trace (B3, B5, B6, B12)	Energy metabolism support
Flavor	As needed	Lemon–berry or citrus
Osmolality	230–260 mOsm/kg	Functional rehydration with added vitamins

8.2 Performance Beverage (Isotonic System)

Performance beverages represent the core of sports hydration science, where fluid, energy, and electrolytes converge to sustain endurance and output. These isotonic systems, balanced to match the body's natural osmolality (~280–310 mOsm/kg), enable simultaneous hydration and carbohydrate energy delivery. This section explores the physiological rationale behind isotonic formulations and how they maintain plasma volume, delay fatigue, and support thermoregulation under physical stress. It guides formulators through optimizing carbohydrate blends for quick and sustained energy, balancing electrolyte ratios to match sweat profiles, and fine-tuning sensory attributes like sweetness and acidity. Readers will also discover how to design isotonic beverages for distinct activity intensities, from team sports to long-distance training, while aligning taste, functionality, and brand positioning for the active lifestyle market.

8.2.1 Classic Sports Isotonic Drink

The classic sports isotonic drink is the foundational model for all modern performance beverages, an optimized mix of carbohydrates and electrolytes that replenishes fluids, restores sodium and potassium, and fuels the body during sustained exertion. This section examines the formulation principles that make isotonic beverages effective: maintaining plasma osmolality, preventing dehydration, and supporting muscular endurance. Readers will gain insight into selecting carbohydrate sources like glucose and maltodextrin for immediate and extended energy, balancing sodium and potassium ratios to match sweat composition, and fine-tuning sensory properties for maximum drinkability. As a proven and trusted format, the classic isotonic drink remains a template for reliable hydration under real-world performance conditions.

Table 8.2.1: Classic Sports Isotonic Drink

Component	Typical Level (per 1 L)	Function / Rationale
Water	q.s.	Base
Sodium chloride	0.9 g (~360 mg Na^+)	Sweat replacement
Potassium citrate	0.35 g	Electrolyte balance
Maltodextrin + glucose	6%	Quick and sustained fuel
Citric acid	0.2%	Flavor and stability
Flavor (orange, lemon)	0.1%	Consumer appeal
Osmolality	290 mOsm/kg	Standard isotonic profile

8.2.2 Low-Sugar Performance Drink

The low-sugar performance drink responds to the evolving demand for cleaner, lighter, and calorie-conscious hydration solutions. By reducing sugar concentration while maintaining isotonic balance, this formulation delivers the hydration benefits of traditional sports drinks without the heavy sweetness or high calorie content. This section explores how to replace or reduce carbohydrates using natural sweeteners, complex carbs, or low-glycemic blends while preserving osmotic efficiency and palatability. It also addresses challenges such as flavor masking, aftertaste control, and energy perception in reduced-sugar systems. Readers will learn how to engineer isotonic beverages that align with modern nutrition trends, offering sustained hydration performance with metabolic balance and clean-label appeal.

Table 8.2.2: Low-Sugar Performance Drink – Components and Functions

Component	Typical Level (per 1 L)	Function / Rationale
Water	q.s.	Base
Sodium chloride	0.7 g	Rehydration
Potassium phosphate	0.3 g	Improves buffering capacity
Maltodextrin	4%	Controlled energy release
Stevia	0.02%	Low-calorie sweetness
Citric acid	0.15%	Freshness
Natural berry flavor	As needed	Light taste
Osmolality	270–290 mOsm/kg	Balanced for endurance without high sugar

8.2.3 Caffeine-Enhanced Energy Hydration

The caffeine-enhanced energy hydration concept merges hydration science with mental and physical stimulation. Designed for high-intensity or endurance activity, these beverages combine isotonic electrolyte systems with moderate caffeine doses and functional co-actives such as taurine or B-vitamins. This section explores how caffeine influences alertness, endurance, and perceived effort, and how to integrate it effectively without compromising hydration efficiency or causing overstimulation. Readers will gain insights into flavor pairing, caffeine stabilization, and regulatory considerations when formulating beverages positioned between energy drinks and sports hydration products. This hybrid model delivers focused energy and fluid balance simultaneously, appealing to consumers seeking sharper performance, faster recovery, and mental clarity under fatigue.

Table 8.2.3: Caffeine-Enhanced Energy Hydration

Component	Typical Level (per 1 L)	Function / Rationale
Water	q.s.	Base
Sodium chloride	0.85 g	Hydration support
Potassium citrate	0.3 g	Electrolyte replacement
Glucose + maltodextrin	6.5%	Energy substrate
Caffeine	50 mg	Alertness and fatigue reduction
Taurine	0.3%	Cognitive and muscular support
Flavor (citrus or tropical)	0.1%	Taste and identity
Osmolality	300 mOsm/kg	Hydration–stimulation hybrid

8.2.4 Natural Fruit Isotonic Beverage

The natural fruit isotonic beverage embodies the intersection of performance and natural nutrition. By utilizing real fruit juices as a carbohydrate and flavor source, these formulations achieve isotonic balance while enhancing antioxidant value, taste authenticity, and visual appeal. This section explores how fruit-derived electrolytes, organic acids, and natural sugars can replace synthetic components while maintaining precise hydration performance. It also covers formulation challenges such as juice acidity, microbial stability, and color consistency. Readers will learn how to design fruit-based isotonic drinks that cater to both active and lifestyle markets, offering the functionality of a sports beverage with the wholesomeness and sensory richness of natural juice hydration.

Table 8.2.4: Natural Fruit Isotonic Beverage

Component	Typical Level (per 1 L)	Function / Rationale
Water	800 mL	Dilution medium
Fruit juice (orange, apple)	200 mL	Natural carbs and flavor
Sea salt	0.8 g	Sodium source
Potassium citrate	0.3 g	Muscle function
Magnesium sulfate	0.05 g	Electrolyte support
Osmolality	290–310 mOsm/kg	Natural isotonic fruit-based product

8.3 Recovery Beverage (Hypertonic System)

After intense exertion, the body's priority shifts from endurance to recovery, replenishing glycogen, repairing muscle tissue, and restoring electrolyte equilibrium. Hypertonic beverages, with solute concentrations higher than plasma, are specifically engineered to support these processes. This section explains the science of post-exercise recovery and how carbohydrate-rich, nutrient-dense systems accelerate glycogen resynthesis and hydration restoration. It covers the integration of proteins, peptides, and amino acids for muscle repair, along with key micronutrients like magnesium and calcium for neuromuscular recovery. Readers will learn to design hypertonic beverages that deliver measurable recovery benefits while maintaining palatable sweetness and stable formulation characteristics. By combining energy density with physiological purpose, these drinks serve as the final step in a complete performance hydration system.

8.3.1 Carbohydrate Recovery Formula

The carbohydrate recovery formula represents the simplest and most efficient post-exercise system for rapid energy restoration. Built around a high-carbohydrate, hypertonic base, it is designed to replenish glycogen stores depleted during training or competition. This section examines the importance of carbohydrate concentration and source, comparing glucose, maltodextrin, and fructose blends for optimized absorption and minimal gastric discomfort. It also discusses the role of electrolytes in restoring plasma volume and promoting rehydration during recovery. As the most fundamental recovery beverage type, the carbohydrate formula demonstrates how to balance energy density and fluid intake for fast and effective post-exercise refueling.

Table 8.3.1: Carbohydrate Recovery Formula

Component	Typical Level (per 1 L)	Function / Rationale
Water	q.s.	Base
Sodium chloride	1.0 g	Sodium replacement
Potassium citrate	0.5 g	Cellular recovery
Maltodextrin + fructose	10%	Glycogen replenishment
Citric acid	0.2%	pH control
Flavor	As needed	Palatability
Osmolality	350 mOsm/kg	Basic high-carb recovery system

8.3.2 Protein–Carbohydrate Recovery Shake

The protein–carbohydrate recovery shake builds upon the carbohydrate base by integrating high-quality protein or hydrolyzed peptides to support muscle synthesis and repair. This formulation is ideal for strength and endurance athletes requiring simultaneous glycogen replenishment and muscle recovery. The section explores how the carbohydrate-to-protein ratio (typically 3:1 or 4:1) affects recovery efficiency, how to select protein sources (whey, casein, or plant-based isolates), and how to maintain palatability and stability in mixed nutrient systems. Readers will also gain practical knowledge about emulsification, flavor optimization, and pH management to prevent precipitation or off-flavors. The result is a multi-functional recovery beverage that bridges sports nutrition and post-performance wellness.

Table 8.3.2: Protein–Carbohydrate Recovery Shake

Component	Typical Level (per 1 L)	Function / Rationale
Water	q.s.	Base
Maltodextrin	8%	Energy and glycogen replenishment
Whey protein isolate	2%	Muscle repair
Sodium chloride	0.9 g	Electrolyte restoration
Potassium phosphate	0.4 g	Buffering and muscle support
Cocoa or vanilla flavor	To taste	Smooth recovery taste
Osmolality	370 mOsm/kg	Balanced energy–protein recovery blend

8.3.3 BCAA-Enriched Endurance Recovery Drink

The BCAA-enriched recovery drink targets endurance athletes seeking faster muscular recovery and reduced fatigue through targeted amino acid supplementation. By combining branched-chain amino acids (leucine, isoleucine, and valine) with a carbohydrate-electrolyte base, this beverage supports protein synthesis and limits muscle breakdown after prolonged exertion. This section discusses how BCAAs interact with energy metabolism, optimal inclusion levels for efficacy, and strategies to mask bitterness while preserving clarity and flavor. Readers will learn formulation techniques for incorporating free amino acids into clear or flavored systems without compromising stability. The BCAA-enriched model showcases how precision nutrition can be seamlessly integrated into a hydration format, creating drinks that restore both energy and performance capacity.

Table 8.3.3: BCAA-Enriched Endurance Recovery Drink

Component	Typical Level (per 1 L)	Function / Rationale

Water	q.s.	Base
Maltodextrin + glucose	9%	Energy supply
BCAA blend (2:1:1)	1%	Reduces muscle breakdown
Sodium chloride	1 g	Hydration
Potassium citrate	0.4 g	Muscle balance
Magnesium lactate	0.1 g	Energy metabolism
Osmolality	360–380 mOsm/kg	Advanced endurance recovery formula

8.3.4 Plant-Based Recovery Beverage

The plant-based recovery beverage reflects the modern shift toward natural, sustainable, and allergen-friendly performance nutrition. Using plant-derived proteins, coconut water, and natural carbohydrate sources, these formulations deliver full recovery functionality without dairy or artificial ingredients. This section explores ingredient sourcing, solubility challenges of plant proteins, and how natural sugars and minerals can mimic traditional hypertonic performance. It also highlights flavor and texture balancing strategies, critical for achieving smooth mouthfeel and pleasant sensory experience in plant systems. Readers will discover how to craft recovery beverages that align with vegan, clean-label, and eco-conscious markets, offering nutrient-rich replenishment grounded in natural authenticity and modern performance science.

Table 8.3.4: Plant-Based Recovery Beverage

Component	**Typical Level (per 1 L)**	**Function / Rationale**
Coconut water	800 mL	Natural electrolyte base
Oat extract or rice protein	2%	Plant protein source

Cane sugar	5%	Natural carbohydrate source
Sea salt	0.8 g	Sodium
Potassium citrate	0.3 g	Potassium source
Cacao or coffee extract	Trace	Flavor and antioxidant support
Osmolality	340 mOsm/kg	Vegan-friendly recovery beverage

8.4 Functional Hydration Beverage (Clean-Label Hybrid)

Functional hydration beverages represent the evolution of modern hydration science, merging physiological performance with consumer demand for natural, transparent, and wellness-oriented products. These clean-label hybrids go beyond traditional sports drinks, combining electrolytes with natural minerals, plant extracts, vitamins, or adaptogens to deliver hydration plus functional enhancement. This section explores how to formulate beverages that hydrate efficiently while aligning with current trends in low-calorie, additive-free, and sustainability-focused product design. From coconut-water-based systems to mineralized functional waters and botanical electrolyte blends, readers will discover formulation strategies that balance efficacy with natural identity. The discussion also highlights labeling, flavor design, and sensory differentiation, key factors that define this rapidly expanding segment bridging performance and lifestyle hydration.

8.4.1 Coconut Water Electrolyte Drink

Coconut water has become a modern hydration icon, valued for its naturally occurring electrolytes and clean-label appeal. When used as the foundation of an electrolyte beverage, it offers a distinctive balance of potassium, mild sweetness, and a refreshing flavor profile that aligns with both performance and wellness markets. This section explores how to elevate coconut water beyond its raw form, optimizing clarity, mineral balance, and stability, while preserving the natural cues consumers expect. Here, formulation science meets a widely trusted ingredient, providing developers with a path to create a premium, nature-forward hydration system anchored in coconut water's intrinsic functional strengths.

Table 8.4.1: Formulation Composition for Coconut Water Electrolyte Drink

Component	Typical Level (per 1 L)	Function / Rationale
Coconut water (single strength)	1000 mL	Natural hydration base
Sea salt	0.4 g	Sodium
Lime juice	2%	Flavor, vitamin C
Osmolality:	240 mOsm/kg	100% natural isotonic-like product

8.4.2 Herbal Hydration Infusion

Herbal hydration infusions bring a botanical dimension to functional beverages, blending light electrolytes with plant-derived actives that support relaxation, focus, or antioxidant benefits. These systems appeal to consumers seeking hydration with a wellness narrative, clean, refreshing, and naturally expressive. This section explores how to integrate herbs, florals, and gentle botanicals into stable, clear beverages while navigating challenges such as polyphenol bitterness, color shifts, and pH sensitivity. By combining purposeful hydration with botanical nuance, herbal infusions offer a path to products that feel both functional and elevated, meeting the growing demand for nature-aligned, mindful refreshment.

Table 8.4.2: Herbal Hydration Infusion

Component	Typical Level (per 1 L)	Function / Rationale

Water	q.s.	Base
Electrolyte blend	300 mg Na^+ / 100 mg K^+	Fluid balance
Green tea extract	0.05%	Antioxidant, mild caffeine
Lemon balm or ginseng	0.02%	Adaptogenic support
Honey	2%	Natural carbohydrate
Lemon flavor	To taste	Fresh sensory cue
Osmolality	230 mOsm/kg	Natural wellness hydration system

8.4.3 Vitamin–Mineral Enhanced Hydration

Vitamin–mineral hydration systems bridge everyday wellness with purposeful functionality, offering consumers a clean and convenient way to support energy metabolism, immunity, and daily micronutrient needs. Designed to balance clarity, stability, and drinkability, these formulations extend hydration beyond electrolytes by integrating essential vitamins and gentle mineral levels without overwhelming taste or osmolality. This section examines how to design nutrient-fortified beverages that deliver meaningful benefits while maintaining refreshing sensory qualities. Here, hydration becomes a multi-benefit platform, precise, approachable, and aligned with the growing demand for smarter, daily functional drinks.

Table 8.4.3: Vitamin–Mineral Enhanced Hydration

Component	Typical Level (per 1 L)	Function / Rationale
Water	q.s.	Base

Sodium chloride	0.6 g	Hydration
Potassium citrate	0.25 g	Muscle recovery
Magnesium sulfate	0.05 g	Metabolism
Vitamin C	80 mg	Antioxidant
Niacin, B6, B12	Trace	Energy metabolism
Stevia	0.02%	Sweetness
Osmolality	240 mOsm/kg	Daily hydration with vitamin support

8.4.4 Electrolyte Sparkling Water

Electrolyte sparkling water blends the purity of lightly flavored carbonated water with the functional boost of essential minerals, creating a refreshing alternative to both traditional sports drinks and standard seltzers. This category appeals to consumers seeking hydration with zero heaviness, clean, crisp, and modern. Formulating these beverages requires careful control of carbonation, acidity, and mineral balance to maintain clarity and a smooth, mineral-forward profile without bitterness. This section explores how to craft sparkling electrolyte systems that feel elevated yet simple, delivering everyday hydration in a format that is contemporary, versatile, and increasingly in demand.

Table 8.4.4: Electrolyte Sparkling Water

Component	Typical Level (per 1 L)	Function / Rationale
Carbonated water	q.s.	Effervescent base

Sodium bicarbonate	0.4 g	Sodium + alkalinity
Potassium chloride	0.25 g	Electrolyte balance
Natural fruit extract	0.1%	Flavor
Stevia or monk fruit	To taste	Sweetness
Osmolality	220 mOsm/kg	Functional sparkling hydration beverage

9. Electrolyte Systems: Building the Core Matrix

This chapter focuses on designing the electrolyte backbone of a performance beverage. It covers how to select and combine salts, balance ionic ratios, and minimize off-flavors associated with high mineral loads. Practical strategies are offered for adjusting electrolyte blends according to sweat rates, activity duration, and intended use, whether for light fitness, endurance sports, or recovery hydration.

9.1 Choosing electrolyte salts and ratios

At the heart of every high-performance hydration beverage lies its electrolyte system, the precise balance of sodium, potassium, magnesium, and calcium that sustains nerve function, muscle contraction, and fluid retention. This section examines the critical process of selecting electrolyte salts based on solubility, bioavailability, taste profile, and compatibility with other ingredients. Sodium chloride, potassium citrate, magnesium lactate, and calcium gluconate are among the most widely used, but each has distinct sensory and functional characteristics that affect both product performance and consumer acceptance. Readers will learn how to calculate ionic ratios that mirror physiological needs, optimize sodium–potassium balance for different sports intensities, and maintain electrolyte stability across varied pH and storage conditions. The goal is to build a core mineral matrix that delivers measurable hydration efficiency without compromising sensory appeal or product integrity.

9.1.1 Solubility and Ionic Dissociation

High solubility ensures rapid dissolution and uniform ion distribution within the beverage, which is essential for maintaining consistent electrolyte delivery and clarity. Sodium and potassium chlorides, for example, are highly soluble and efficient conductors of hydration, but they can create harsh saltiness if used excessively. Citrate and lactate salts offer slightly lower solubility but improved palatability, making them suitable alternatives in flavored beverages. Magnesium and calcium salts, particularly their carbonate and sulfate forms, tend to exhibit lower solubility and can cause haze or sedimentation if not properly managed. To address this, formulators often use organic acid salts such as magnesium lactate or calcium gluconate, which dissolve more completely and remain stable across a range of pH values.

9.1.2 Bioavailability and Physiological Function

Not all salts deliver minerals equally, bioavailability depends on how efficiently ions are absorbed and utilized by the body. Sodium chloride remains the most efficient for plasma volume maintenance, while potassium citrate and magnesium lactate provide superior cellular uptake due to their organic anion partners. Citrate-based salts have the added benefit of buffering lactic acid accumulation, helping delay fatigue during intense exercise. Calcium gluconate and calcium lactate are preferred for their high absorption and mild taste, while phosphates can enhance energy metabolism but may raise formulation pH and cloud clarity. A well-designed electrolyte matrix therefore uses a multi-salt approach, balancing immediate plasma hydration with sustained cellular support.

9.1.3 Taste Profile and Sensory Impact

Electrolytes are functionally vital but sensorially challenging. Chloride salts deliver sharp saltiness; magnesium and calcium can introduce bitterness, metallic notes, or chalky mouthfeel. The choice of salt form greatly influences flavor perception, citric, malic, and lactate salts produce softer, more rounded taste profiles. Potassium citrate, for example, contributes gentle tartness that complements fruit-based flavors, while magnesium lactate offers a cleaner mineral finish compared to magnesium chloride. Formulators often use flavor modulation, pairing electrolytes with acids, sweeteners, or masking agents, to maintain the balance between authentic taste and perceived freshness. Sensory testing at various ionic strengths is essential, as the interaction between electrolytes, acids, and carbohydrates changes both perceived sweetness and mouthfeel.

9.1.4 Compatibility with Other Ingredients

Electrolyte salts must remain chemically stable and non-reactive within complex beverage systems containing carbohydrates, vitamins, amino acids, and natural flavors. High sodium and magnesium levels can destabilize emulsions or precipitate with phosphates and proteins. Similarly, certain flavor compounds or acidulants can alter ionic strength, impacting clarity and osmolality. Choosing compatible anions, such as citrate or gluconate, helps minimize these risks by maintaining neutral pH and preventing ionic cross-reactions. For example, calcium chloride may interact with pectin or protein-based ingredients, causing haze or sedimentation, whereas calcium lactate avoids such interactions while delivering equivalent functionality.

9.1.5 Comparative Properties of Common Electrolyte Salts

Electrolyte salts may appear interchangeable on a label, but their functional behavior in a beverage system can differ dramatically. Solubility, taste impact, ionic strength, buffering effects, and interactions with acids or sweeteners all influence how a hydration product performs, and how consumers perceive it. This section breaks down the practical distinctions between the most commonly used electrolyte salts, helping formulators choose the right form for each application. By understanding these nuanced differences, developers can fine-tune flavor, stability, cost, and performance with far greater precision, ensuring the beverage delivers on both functionality and drinking experience.

Table 9.1.5: Comparative Properties of Common Electrolyte Salts for Beverage Formulation

Electrolyte Salt	Primary Ion(s) Delivered	Relative Solubility (in Water)	Bioavailability / Absorption Efficiency	Taste Profile	Functional Advantages	Formulation Considerations
Sodium Chloride (NaCl)	Na^+, Cl^-	Very high (~36	Excellent, rapidly absorbed	Strongly salty; sharp	Fast sodium replacement;	Overuse increases saltiness; may mask

		g/100 mL)			maintains plasma volume	delicate flavors
Sodium Citrate ($Na_3C_6H_5O_7$)	Na^+	Moderate (~9 g/100 mL)	High; citrate improves buffering and absorption	Mild, slightly tart	Buffers acidity; smooths flavor; aids pH stability	Can raise pH slightly; adds tartness in low-acid systems
Sodium Bicarbonate ($NaHCO_3$)	Na^+	Moderate (~10 g/100 mL)	High	Mildly salty–alkaline	Adds alkalinity; neutralizes acidity; aids CO_2 retention in sparkling drinks	Can cause CO_2 release or foaming; affects carbonation
Potassium Chloride (KCl)	K^+, Cl^-	High (~34 g/100 mL)	Good	Bitter, metallic aftertaste	Replaces potassium; supports muscle and nerve function	Bitterness requires masking; often blended with citrate forms
Potassium Citrate ($K_3C_6H_5O_7$)	K^+	Moderate (~20 g/100 mL)	Excellent	Mildly salty–tart; smooth finish	Balances intracellular fluids; buffers lactic acid	Mildly acidic; complements fruit-based

						flavor systems
Magnesium Chloride ($MgCl_2$)	Mg^{2+}, Cl^-	High (~54 g/100 mL)	Moderate	Bitter, metallic	Effective for magnesium delivery; small doses only	Strong flavor impact; limit inclusion to <0.05%
Magnesium Lactate ($Mg(C_3H_5O_3)_2$)	Mg^{2+}	Moderate (~10 g/100 mL)	High	Milder mineral taste; slightly sweet	High bioavailability; gentle on stomach	Costlier; sensitive to pH <3.5 (may form haze)
Magnesium Sulfate ($MgSO_4$)	Mg^{2+}	Moderate (~35 g/100 mL)	Moderate	Bitter, astringent	Adds electrolyte density; aids muscle relaxation	Taste masking required; high sulfate may impact clarity
Calcium Lactate ($Ca(C_3H_5O_3)_2$)	Ca^{2+}	Moderate (~9 g/100 mL)	High	Mild, neutral	Excellent calcium source; compatible with fruit acids	Slight turbidity in low-acid beverages
Calcium Gluconate ($C_{12}H_{22}CaO_{14}$)	Ca^{2+}	Moderate (~3.5	High	Very mild; neutral	Stable, non-reactive	Lower solubility; may require

		g/100 mL)			calcium source	pre-dissolution
Trisodium Phosphate (Na_3PO_4)	Na^+, PO_4^{3-}	High (~14 g/100 mL)	Good	Mildly salty	Enhances buffering and energy metabolism	May raise pH; affects clarity and flavor in low-acid systems

Formulation Insight

- Citrate and lactate salts are preferred for palatable, consumer-facing beverages, offering smooth taste and buffering properties.
- Chloride salts provide strong electrolyte function but need flavor balancing to avoid harshness.
- Organic acid salts (lactates, gluconates) enhance bioavailability and are well-suited to premium or clean-label hydration systems.
- Balanced multi-salt systems, combining sodium chloride, potassium citrate, and magnesium lactate, often deliver the best compromise between function, flavor, and stability.

9.2 Managing taste (bitterness, saltiness) and mouthfeel

Electrolyte systems are central to functional hydration, yet they introduce one of the greatest formulation challenges: achieving a pleasant sensory experience despite high mineral loads. While sodium, potassium, magnesium, and calcium are physiologically indispensable, they impart distinct flavor, aroma, and mouthfeel effects that can diminish consumer acceptance if not carefully balanced. This section explores advanced formulation strategies for managing these sensory attributes, transforming technically potent electrolyte matrices into beverages that are clean, smooth, and refreshingly drinkable.

9.2.1 Reducing Harshness through Salt Selection

Different salt forms carry dramatically different sensory impacts. Chloride salts, while effective, tend to introduce sharp saltiness and metallic edges, especially at concentrations above 0.5%. Replacing part or all of these with organic acid salts, such as citrate, lactate, or gluconate, softens the overall taste profile and improves flavor integration. Sodium citrate and potassium citrate provide gentle tartness and natural buffering, while magnesium lactate delivers a milder mineral note compared to the harsh bitterness of magnesium chloride. By combining salts with complementary flavor characteristics, formulators can maintain ionic balance while minimizing bitterness and salinity.

9.2.2 Flavor-Masking and Sweetness Optimization

Flavor-masking technologies are invaluable when dealing with persistent mineral notes or metallic aftertastes. Common strategies include using sweet–acid synergy, where acids (citric, malic, or ascorbic) sharpen sweetness perception and suppress bitterness, or natural flavor modulators such as glycyrrhizic acid (licorice extract) or monk fruit derivatives that smooth taste transitions. Sweetener choice also plays a role: sucrose offers ideal roundness but raises osmolality, while stevia and sucralose provide calorie-free sweetness with potential lingering notes that must be balanced by acidity and aroma. Achieving a harmonious acid–sweetener ratio is key, too much acid may accentuate bitterness, while excessive sweetness can mask freshness and reduce perceived hydration.

9.2.3 Layering Acidity, Flavor, and Aroma

Effective sensory design goes beyond masking, it's about layering flavors to achieve natural complexity. Acidity defines brightness and thirst-quenching character; fruit esters and aldehydes provide aroma cues that psychologically reinforce freshness; and small additions of salt itself can enhance perceived sweetness through contrast. For example, a citrus-flavored isotonic beverage benefits from pairing potassium citrate (for tartness) with malic acid and lemon-lime top notes. This layering not only conceals off-flavors but also creates a more multidimensional drinking experience, encouraging repeat consumption under physical stress or fatigue.

9.2.4 Optimizing Mouthfeel and Drink Weight

Electrolytes influence not only taste but also mouthfeel and body perception. High ionic strength can increase viscosity slightly, creating a thicker or heavier sensation that may feel less refreshing. Balancing osmolality and acid–carbohydrate composition helps control drink "weight." Light hypotonic formulations should feel crisp and clean, while isotonic or recovery beverages may benefit from a slightly fuller mouthfeel associated with energy and replenishment. Ingredients such as low-level hydrocolloids, glycerol esters, or natural polysaccharides (e.g., from aloe or pectin) can subtly modify body without adding calories or turbidity.

9.2.5 Integrating Sensory Science into Formulation Design

Developing electrolyte beverages that are scientifically functional yet sensorially rewarding requires integrating flavor chemistry with nutrition science. Sensory mapping, tracking how sodium, potassium, acids, and sweeteners interact, helps identify the optimal balance between perceived hydration, sweetness, and freshness. Regular panel testing at realistic serving temperatures (5–10°C) provides critical data on flavor release and mouthfeel under consumption conditions.

Sensory Impact of Common Electrolyte Salts and Flavor Pairing

The following table summarizes the sensory behavior of commonly used electrolyte salts, providing quick formulation insight into their flavor character, perceived intensity, and masking or pairing strategies, invaluable for beverage developers balancing performance with drinkability.

Table 9.2.5A: Sensory Impact of Common Electrolyte Salts and Flavor Pairing Strategies

Electrolyte Salt	**Primary Ion(s)**	**Taste Description**	**Relative Taste Intensity (1–5)**	**Aftertaste / Mouthfeel Effect**	**Ideal Flavor Pairings / Masking Strategies**	**Formulation Notes**

Sodium Chloride (NaCl)	Na^+, Cl^-	Clean salty, sharp, enhances overall flavor	3	Crisp, slightly drying	Pairs well with citrus, lemon-lime, and berry flavors; enhances sweetness perception	Use below 0.1% to round flavor; excess leads to harsh saltiness
Sodium Citrate ($Na_3C_6H_5O_7$)	Na^+	Mildly salty-tart, buffering acidity	2	Smooth, refreshing	Excellent with citrus, apple, and tropical blends; adds brightness	Improves flavor balance and stability in fruit-based drinks
Potassium Chloride (KCl)	K^+, Cl^-	Salty-bitter, slightly metallic	4	Lingering bitterness	Best masked with strong acids (citric, malic) or dark fruit (grape, pomegranate)	Use up to 0.3–0.4 g/L; combine with sweetener and acid balance
Potassium Citrate ($K_3C_6H_5O_7$)	K^+	Mildly tart, soft mineral note	2	Smooth and light	Ideal for citrus and tropical profiles;	Preferred over KCl for taste and

					complements mango, orange, pineapple	buffering properties
Magnesium Chloride ($MgCl_2$)	Mg^{2+}, Cl^-	Bitter, metallic, earthy	5	Lingering dryness	Mask with bold citrus (lemon, grapefruit) or flavor enhancers like ginger	Use minimal levels (<0.05%); combine with lactate or citrate forms
Magnesium Lactate ($Mg(C_3H_5O_3)_2$)	Mg^{2+}	Mild, slightly sweet mineral	2	Clean finish	Works with citrus, melon, and berry; subtle in low-acid beverages	Highly bioavailable and gentle; ideal for premium formulations
Magnesium Sulfate ($MgSO_4$)	Mg^{2+}, SO_4^{2-}	Bitter, astringent	4	Chalky, dry finish	Mask with strong acid–sweet combinations; suitable for dark fruit or cola-type bases	Avoid high inclusion; can dull delicate fruit notes

Calcium Lactate ($Ca(C_3H_5O_3)_2$)	Ca^{2+}	Neutral, slightly milky	1	Smooth, round	Compatible with creamy, tropical, or vanilla-based flavors	Provides body and mouthfeel; improves texture in recovery beverages
Calcium Gluconate ($C_{12}H_{22}CaO_{14}$)	Ca^{2+}	Very mild, neutral	1	Soft, minimal	Blends with any flavor; best for clear, delicate beverages	Low solubility, pre-dissolve or use in micro-encapsulated form
Sodium Bicarbonate ($NaHCO_3$)	Na^+	Mildly salty–alkaline	2	Slight effervescence	Suits citrus, mint, and sparkling formats; enhances freshness	Buffers acidity; supports carbonation stability
Trisodium Phosphate (Na_3PO_4)	Na^+, PO_4^{3-}	Mildly saline–soapy	3	Coats mouth, light film	Works with strong acids, cola, or tropical fruit	Use in low levels; excessive phosphate raises pH and dulls flavor

Sea Salt Extract / Natural Mineral Blend	Na^+, K^+, Mg^{2+} (trace Ca^{2+})	Balanced mineral complexity; slight brine	3	Natural minerality, smooth	Pairs well with citrus, cucumber, or coconut bases	Excellent for "natural electrolyte" and clean-label positioning

Interpretation and Application Guidelines

- Chloride salts provide authentic electrolyte function but need acid–sweetness pairing for palatability.
- Citrate and lactate salts deliver smoother, rounder flavor and double as buffering agents to stabilize acidity.
- Magnesium and potassium salts contribute functional value but require flavor masking or dilution to avoid metallic or bitter notes.
- Calcium salts add subtle body and smoothness, beneficial for recovery and hybrid performance beverages.
- Combining acids (citric, malic, tartaric) with natural fruit essences creates a refreshing flavor architecture that enhances perceived hydration.

Integrating Mouthfeel Modifiers and Sensory Enhancers in Electrolyte:

Below table summarizes commonly used mouthfeel modifiers, sensory enhancers, and texturizing agents, detailing their functional roles, usage levels, and interactions with electrolyte systems. It's tailored for both performance-oriented and clean-label formulations.

Table 9.2.5B: Mouthfeel Modifiers and Sensory Enhancers in Electrolyte and Hydration Beverages

Ingredient / Additive	Functional Category	Typical Usage Range (% w/v)	Primary Sensory Effect	Mechanism of Action	Ideal Application Context	Compatibility / Notes
Glycerol (Glycerin)	Humectant / Texture Enhancer	0.2–0.8%	Adds body and slight sweetness; smooth, round mouthfeel	Increases viscosity and water-binding capacity	Isotonic and recovery beverages requiring fuller mouthfeel	Enhances perceived hydration; avoid >1% to prevent syrupy sensation
Pectin (Low-Methoxyl)	Natural polysaccharide	0.05–0.2%	Smooth texture, mild thickness, improved mouth coating	Forms weak gels and increases viscosity at low pH	Recovery and fruit-based isotonic beverages	Compatible with calcium; adjust levels to prevent haze
Xanthan Gum	Hydrocolloid / Stabilizer	0.02–0.1%	Adds slight viscosity and cling; improves body	Creates shear-thinning texture; stabilizes particulates	Functional and high-electrolyte systems needing	Blend with sucrose or gums for clarity; avoid excess for watery drinks

					visual stability	
Acacia Gum (Gum Arabic)	Natural emulsifier / Texturizer	0.05–0.3%	Clean, soft mouthfeel; stabilizes flavor oils	Provides colloidal stability and flavor uniformity	Clear isotonic beverages and citrus-flavored systems	Excellent clarity; low viscosity impact
Gellan Gum (Low-Acyl)	Structuring agent / Suspension aid	0.02–0.06 %	Lightly structured, crisp texture; maintains suspended minerals or particles	Forms transparent fluid gels	Premium hydration waters or beverages with visible ingredients	Works best in low pH; synergistic with calcium ions
Guar Gum (Clarified)	Thickener / Mouth-coating enhancer	0.03–0.1%	Adds soft viscosity and rounded texture	Hydrophilic polymer increases bulk and retention	Plant-based recovery or meal-replacement beverages	Can reduce clarity; not ideal for transparent formulations
Beta-Glucans (Oat or	Functional fiber /	0.05–0.2%	Adds creaminess, mild thickness	Forms soluble fiber matrix increasing	Recovery and hybrid hydratio	Contributes fiber claim; slightly

Barley Extract)	Natural thickener		, and satiety feel	drink weight	n-protein beverages	dulls fruit brightness
Aloe Vera Extract (De-colorized)	Botanical polysaccharide	0.1–0.5%	Smooth, hydrating texture; perceived freshness	Polysaccharides enhance lubricity and cool sensation	Functional "wellness hydration" drinks	Supports clean-label positioning; limited stability in low pH
Polydextrose	Soluble fiber / Bulk sweetener	0.5–2.0%	Adds mild viscosity and sweetness balance	Mimics sugar mouthfeel; low-calorie bulking	Low-sugar isotonic beverages	Neutral taste; compatible with most electrolytes
Sorbitol or Erythritol	Polyol / Cooling agent	0.3–1.0%	Cooling effect, enhances refreshment	Endothermic dissolution; modifies sweetness profile	Clean-label hypotonic hydration and wellness waters	Avoid high levels to prevent laxative effect
Hydrolyzed Collagen Peptides	Protein-based texturizer	0.5–2.0%	Adds body, smoothness, and slight richness	Protein structure increases viscosity and film-forming	Recovery and protein–electrolyte blends	Heat stable; minimal impact on clarity at low inclusion

Coconut Water Concentrate	Natural electrolyte base	2–5%	Natural body, slight sweetness, mineral authenticity	Provides inherent potassium and natural sugars	Natural isotonic and functional hydration beverages	Works synergistically with sea salt; enhances natural taste
Citric and Malic Acids (in low levels)	Flavor–acidity balancer	0.05–0.2%	Increases perceived refreshment; heightens brightness	Stimulates salivation and enhances flavor release	All categories; especially light hydration beverages	Controls sharpness of mineral taste; synergistic with sweetness

Formulation Insights and Best Practices

- Low-viscosity modifiers (e.g., glycerol, acacia gum) are preferred in isotonic and hypotonic beverages to maintain clarity and rapid absorption.
- Hydrocolloids and polysaccharides add smoothness and flavor persistence but must remain below haze thresholds for transparent products.
- Acid–sweetener balance works synergistically with texture modifiers to increase perceived refreshment, especially in citrus or tropical profiles.
- For clean-label positioning, choose natural origin options like pectin, aloe extract, and coconut water, which enhance mouthfeel without synthetic additives.

- The ideal hydration beverage texture is light yet cohesive, visually clear, hydrating on entry, and smooth in after-feel, signaling both freshness and quality.

Acid–Sweetener Synergy Map for Hydration and Performance:

This table outlines how acids and sweeteners interact to shape the perceived sweetness, freshness, and overall sensory balance of electrolyte and performance beverages. It provides a concise formulation guide for selecting acid–sweetener systems that deliver optimal taste brightness, flavor clarity, and reduced bitterness, critical for electrolyte-heavy formulations.

Table 9.2.5C: Acid–Sweetener Synergy Map for Hydration and Performance Beverages

Acid Type	**Typical Usage Range (% w/v)**	**Sensory Character**	**Primary Sweetener Synergy**	**Flavor Impact / Key Effect**	**Best Beverage Applications**	**Formulation Considerations**
Citric Acid	0.08–0.25%	Bright, sharp, refreshing tartness	Sucrose, stevia, sucralose	Amplifies perceived sweetness and freshness; masks saltiness and bitterness	Core acidulant for isotonic and hypotonic citrus beverages	Overuse may produce sour bite; balance with potassium citrate for smoothness
Malic Acid	0.05–0.20%	Smooth, lingering	Sucrose, monk fruit,	Enhances fruit body and	Apple, berry, and tropical-	Provides flavor “roundness”

		tartness; fruity	acesulfame-K	balances metallic potassium or magnesium notes	based hydration or recovery drinks	; synergistic with natural fruit esters
Tartaric Acid	0.02–0.10%	Crisp, astringent, grape-like tartness	Sucrose, sucralose	Provides high-acid "bite" for sharp flavor systems	Sports or endurance drinks with bold fruit notes	Use sparingly; may accentuate mineral harshness in high-sodium systems
Lactic Acid	0.03–0.10%	Soft, dairy-like tang; mild acidity	Sucrose, allulose	Adds smoothness and mild sweetness; improves body and perception of fullness	Recovery and protein–electrolyte beverages	May raise turbidity; stable under low heat
Ascorbic Acid (Vitamin C)	0.02–0.08%	Light, tangy, slightly bitter	Sucrose, stevia	Adds clean brightness and functional antioxida	Functional hydration waters and clean-label sports beverages	Avoid high inclusion in metal-rich systems; may oxidize color/flavor

				nt positioning		
Fumaric Acid	0.02–0.06%	Intense, persistent tartness	Sucralose, acesulfame-K	Boosts high-acid intensity without sourness overload	Citrus or tropical isotonic beverages	Insoluble at high levels; pre-dissolve for clarity
Phosphoric Acid	0.02–0.05%	Dry, cola-like acidity	Sucralose, HFCS, or stevia	Adds depth, structure, and "crispness" to dark or cola-type hydration drinks	Caffeinated performance and energy beverages	Not clean-label; may reduce calcium bioavailability
Acetic Acid (Apple Cider or Fermented Sources)	0.03–0.08%	Tangy, fermented note	Stevia, monk fruit	Adds natural sourness; supports probiotic or fermented beverage claims	Functional wellness or detox hydration beverages	May cause haze or flavor instability; use decolorized sources

Citric + Malic Blend (1:1 to 2:1)	0.08–0.25% combined	Balanced bright and smooth acidity	Any sweetener type	Enhances both sweetness perception and natural fruit authenticity	Most isotonic and clean-label hydration formulas	Industry-standard acid system for balanced sensory experience
Malic + Lactic Blend	0.05–0.15% combined	Rounded acidity with mild creaminess	Stevia, monk fruit	Adds body and smooth transition; minimizes bitterness from electrolytes	Recovery, protein-based, or hybrid hydration drinks	Great for low-calorie systems needing "natural feel"
Citric + Ascorbic System	0.05–0.20% combined	Fresh, zesty, clean	Sucrose, sucralose	Brightens flavor while adding functional antioxidant benefit	Vitamin-enhanced hydration and isotonic beverages	Monitor oxidation in transparent packaging
Phosphoric +	0.05–0.15%	Deep, balanced acidity	Sucralose, stevia	Creates cola-style	Performance energy-	Not suitable for clean-label claims

Citric System	combined	with crisp finish		bite; enhances caffeine-based systems	hydration blends

Formulation Insights

- Citric and malic acids form the foundation of most hydration beverages, providing optimal brightness, fruit authenticity, and masking of salty or bitter notes.
- Combining acids creates depth and dynamic flavor transitions, citric for initial impact, malic or lactic for lingering smoothness.
- Acid–sweetener balance directly affects perceived hydration: too little acidity flattens flavor, while excess reduces drinkability under exertion.
- Low-calorie systems benefit from pairing stevia or sucralose with mild acids like malic or lactic to prevent aftertaste.
- In clean-label or functional beverages, use fruit-derived acids (citric, malic, ascorbic) to reinforce natural perception and nutritional positioning.

This table presents a technical overview of sweetener systems used in hydration, performance, and recovery beverages, including both caloric and non-caloric options. It details sweetness potency, flavor characteristics, and ideal pairing strategies to balance electrolytes, acids, and active ingredients without compromising stability or clean-label appeal.

Sweetener Systems:

Table 9.2.5D: Sweetener Systems for Electrolyte, Performance, and Recovery Beverages

Sweetener Type	**Sweetness Intensity (vs.**	**Onset &**	**Taste Profile**	**Functional / Caloric**	**Ideal Applications**	**Synergy & Pairing**	**Formulation**

	Sucrose = 1.0)	Duration		Value (kcal/g)		Strategies	Considerations
Sucrose	1.0	Fast onset, medium decay	Clean, rounded sweetness	4.0	Isotonic & recovery drinks; traditional sports beverages	Pairs with citric or malic acid; enhances perceived hydration	Raises osmolality; limits low-calorie claims
Glucose (Dextrose)	0.7–0.8	Very fast onset, short duration	Mild, less sweet, "clean"	4.0	Rapid hydration & recovery formulations	Supports osmolality and energy delivery	Ideal for rehydration and recovery blends; may taste flat alone
Fructose	1.2–1.5	Medium onset, long-lasting	Fruity, smooth sweetness	4.0	Fruit-based isotonic beverages	Enhances fruit flavors and mouthfeel	Avoid excess; may cause crystallization or GI distress
Maltodextrin	0.1–0.2	Very slow, neutral	Minimal sweetness, adds body	4.0	Recovery and endurance beverages	Combines with sucrose or stevia to	Increases viscosity and caloric load

						balance texture	
Stevia (Rebaudioside-A or D)	150–300	Slight delay, long finish	Clean to slightly herbal	0	Low-calorie, clean-label hydration & wellness drinks	Excellent with citric/malic acid to reduce aftertaste	Overuse may cause lingering bitterness; blend with erythritol or sucralose
Monk Fruit (Luo Han Guo Extract)	150–200	Smooth onset, clean finish	Mild, fruity sweetness	0	Natural low-calorie and functional hydration beverages	Pairs well with stevia or fructose for flavor balance	Heat stable; mild color tint at high inclusion
Sucralose	600	Fast onset, long clean finish	Very sweet, sugar-like	0	Zero-calorie isotonic & performance beverages	Works synergistically with acesulfame-K for depth	Highly stable in heat, pH, and light; overuse can suppress flavor notes

Acesulfame-K (Ace-K)	200	Fast onset, short finish	Sharp, sweet–slightly bitter	0	Energy-hydration blends; isotonic low-calorie	Best in 5–20% blend with sucralose or stevia	Adds “top-note” sweetness; may accentuate acidity
Aspartame	180–200	Medium onset, short finish	Clean, rounded sweetness	4.0	Low-calorie isotonic and recovery beverages	Blends with citric acid for balanced taste	Not heat stable; avoid pasteurization systems
Erythritol	0.6–0.7	Fast onset, cool finish	Clean, mildly cooling	0.2	Clean-label hydration and functional waters	Pairs with stevia or monk fruit for natural sweetness	Cooling effect enhances refreshment perception
Allulose (D-Psicose)	0.7	Fast onset, short decay	Sugar-like, mild	0.2–0.4	Low-calorie isotonic or recovery systems	Enhances flavor body, reduces bitterness	Excellent for “reduced sugar” claims; slightly affects osmolality
Xylitol	1.0	Fast onset,	Sugar-like,	2.4	Specialty	Blends with	Adds freshness

		coolin g	mild coolin g		function al beverag es	stevia or acesulfa me-K	but may cause GI upset at >2%
Agave Syrup	1.2	Medi um onset, smoot h	Slight caram el–fruit tone	3.8	Natural isotonic or recovery drinks	Works with citric or lactic acid for clean natural flavor	High fructose content; impacts osmolality and labeling
Coconut Sugar / Palm Sugar	0.9	Medi um onset, earthy	Warm, natural sweetn ess	3.8	Clean-label electroly te and natural hydratio n waters	Pair with malic acid and fruit notes	Adds color and minerals; not suitable for clear products
Honey / Fruit Concentr ate Blend	0.9–1.2	Medi um onset, long finish	Natura l, floral	3.5–4.0	Premiu m or function al hydratio n systems	Comple ments citrate and lactic acid bases	Limited clarity; may ferment under poor storage

Formulation Insights & Best Practices

- Blended sweetener systems (e.g., sucralose + Ace-K or stevia + erythritol) deliver more natural, rounded sweetness and mask bitterness from electrolytes.

- Acid–sweetener synergy enhances perception of refreshment: use citric or malic acids to boost brightness and suppress lingering sweetness.
- For clean-label products, natural systems like stevia + monk fruit + allulose achieve balanced taste with minimal off-notes and zero calories.
- Avoid exceeding 350–400 mOsm/kg osmolality in isotonic systems when using caloric sweeteners; excess sugar compromises hydration rate.
- Sweetener combinations should be validated under target pH, temperature, and packaging to ensure long-term stability and consistent flavor release.

Flavor Family Compatibility:

This table maps flavor families to their optimal electrolyte, acid, and sweetener pairings, guiding beverage formulators toward systems that deliver both physiological performance and consumer-pleasing sensory balance. It's tailored for rehydration, performance, and recovery beverage applications, with attention to palate refreshment, stability, and flavor authenticity.

Table 9.2.5E: Flavor Family Compatibility with Electrolyte and Sweetener Systems

Flavor Family	Recommended Electrolyte System	Ideal Acid Profile	Best Sweetener / Combination	Sensory Character & Market Positioning	Formulation Tips & Notes
Citrus (Lemon, Lime, Orange, Grapefruit)	NaCl + K-Citrate + Mg-Lactate	Citric (primary) + Malic (secondary)	Sucrose / Stevia + Erythritol	Bright, sharp, energizing; associated with classic sports and	Use chloride forms for authenticity; high-acid tolerance enhances perceived

				isotonic beverages	refreshment; maintain pH ~3.2–3.8
Berry (Strawberry, Blueberry, Raspberry, Acai)	Na-Citrate + K-Citrate + Mg-Lactate	Malic (primary) + Citric (secondary)	Stevia + Allulose / Sucralose	Sweet–tart balance with rich aroma; ideal for functional or antioxidant hydration	Malic acid supports berry depth; color stability critical under light exposure
Tropical (Mango, Pineapple, Passionfruit, Guava)	NaCl + K-Phosphate + Ca-Lactate	Citric + Ascorbic	Stevia + Monk Fruit / Sucrose	Exotic and refreshing; strong flavor masks mineral notes well	Use natural fruit essences; pair with moderate sweetness; balance acidity to avoid sourness
Apple / Pear	Na-Citrate + K-Citrate	Malic (dominant)	Allulose / Monk Fruit + Stevia	Smooth, crisp, natural; aligns with wellness and low-	Avoid excessive sodium; enhance with subtle green or floral notes

				sugar hydration	for authenticity
Watermelon / Melon	Na-Citrate + K-Lactate + Mg-Lactate	Malic + Citric (low ratio)	Stevia + Erythritol / Sucralose	Light, hydrating, clean; suitable for hypotonic rehydration waters	Maintain low acidity (pH >3.8); balance sweetness to avoid syrupy taste
Coconut / Tropical Water	Sea Salt Extract + K-Citrate + Mg-Lactate	Lactic + Citric (mild)	Allulose / Coconut Sugar (natural)	Mild, natural, mineral freshness; perfect for clean-label isotonic beverages	Add trace calcium for authenticity; low flavor loading maintains natural water character
Cucumber / Mint / Herbal	Na-Citrate + K-Lactate	Citric (light) + Malic	Stevia + Monk Fruit	Crisp, refreshing, spa-style; popular in wellness hydration and premium waters	Use natural oils sparingly; ensure emulsification stability; low sweetness for sophistication

Ginger / Citrus Spice	NaCl + K-Citrate + Mg-Sulfate	Citric + Ascorbic	Sucralose + Ace-K / Stevia	Warming-spicy balance; functional or energy-hydration positioning	Add small capsaicin or extract for depth; stable in heat-treated products
Cola / Dark Fruit (Cherry, Blackcurrant)	NaCl + K-Phosphate + Ca-Lactate	Phosphoric + Citric	Sucralose + Ace-K	Rich, structured acidity; performance or caffeine-based systems	Avoid calcium phosphate precipitation; high-acid profile suits energy blends
Vanilla / Cream / Dairy-Inspired	Na-Citrate + Ca-Lactate + Mg-Lactate	Lactic (primary) + Malic	Sucrose / Allulose	Smooth, full-bodied; ideal for recovery or hybrid nutrition drinks	Use stabilizers (pectin, xanthan) for body; watch pH (>3.8) to avoid curdling
Green Tea / Matcha / Herbal Infusion	Na-Citrate + K-Citrate	Malic + Ascorbic	Stevia + Monk Fruit	Subtle, functional; antioxidant hydration and wellness market	Add polyphenols at low levels; mild acid enhances perceived freshness

Grape / Pomegranate	NaCl + K-Citrate + Mg-Lactate	Tartaric + Malic	Sucralose / Stevia + Erythritol	Deep, fruity, slightly tannic; recovery or antioxidant positioning	Strong natural masking ability for potassium bitterness; monitor color stability
Berry–Citrus Hybrid (Blue Raspberry, Tropical Punch)	NaCl + K-Citrate + Mg-Lactate	Citric + Malic (balanced)	Sucralose + Ace-K / Stevia + Erythritol	Vibrant, youth-oriented, energetic flavor	Ideal for isotonic and performance beverages; high acidity masks mineral notes
Botanical / Floral (Lavender, Hibiscus, Rose)	Sea Salt Extract + K-Lactate	Citric (low) + Lactic	Monk Fruit + Allulose	Delicate, premium sensory appeal; fits functional or wellness categories	Keep acidity low; prioritize clarity; pairs well with natural mineral electrolytes

Formulation Insights & Practical Notes

- Citrus, berry, and tropical families offer the strongest masking effects for salty or bitter electrolytes and are ideal for mainstream isotonic beverages.

- Botanical, herbal, and coconut systems align with premium or clean-label positioning, emphasizing refreshment over intensity.
- Use acid layering (citric + malic or lactic) to enhance depth, while controlling pH for clarity and stability.
- Flavor strength should scale with electrolyte load; high sodium formulations benefit from stronger acids and aromatics.
- For multi-functional beverages (with caffeine, vitamins, or amino acids), choose robust flavor profiles (citrus, berry, cola) to maintain sensory harmony.

Stability and Packaging Compatibility Matrix:

This table provides a practical overview of how formulation components (electrolytes, acids, and flavor systems) interact with packaging materials and storage conditions, ensuring both sensory integrity and physicochemical stability throughout shelf life.
It's written for product developers and process engineers preparing hydration, isotonic, and recovery beverages for commercial-scale filling and distribution.

Table 9.2.5F: Stability and Packaging Compatibility Matrix for Electrolyte and Hydration Beverages

Packaging Material / Type	Typical Beverage Category	Electrolyte Compatibility	Acid / pH Range Compatibility	Flavor & Color Stability	Shelf-Life Performance	Key Advantages	Limitations / Critical Notes
PET (Polyethylene Terephthalate) Bottles	Isotonic, Hypotonic, Functional	Excellent with Na^+, K^+, Mg^{2+}, Ca^{2+} salts	Stable pH 2.8–4.2	Moderate to high; color fade with UV	6–12 months (with light barrier)	Lightweight, low cost, recyclable	Sensitive to heat and UV; avoid high-

	Hydration			exposure			ascorbic or iron systems; use UV-blocked PET for fruit acids
Glass Bottles (Clear or Amber)	Premium Hydration, Recovery, Clean-Label	Universal compatibility	Full pH range (2.5–7.0)	Excellent; preserves volatile flavors and natural colors	12–24 months	Absolute flavor protection; inert and non-reactive	Heavier and costlier; fragile; not ideal for mass-market isotonic lines
Aluminum Cans (Epoxy or BPA-free lining)	Performance & Energy-Hydration	Excellent; stable with all salts if coated	Stable pH 2.8–4.0	High; protects from light and oxygen	9–15 months	Excellent oxygen barrier; protects sensitive vitamin	Internal lacquer essential to prevent ion migration (Mg^{2+}, Na^{+}, Cl^{-});

						s and flavors	avoid unlined cans
Tetra Pak (Aseptic Carton)	Natural & Functional Hydration Waters	Excellent with citrate and lactate salts; avoid chloride forms	pH 3.0–4.5 ideal	Very good; protects light-sensitive nutrients	9–12 months (unopened)	Light and oxygen barrier; ideal for natural-positioned drinks	Not suitable for carbonation; slight flavor absorption over time
HDPE Bottles (Opaque)	Recovery, Nutritional & Protein Hydration	Excellent for Ca^{2+} and Mg^{2+} systems	Stable pH 3.5–6.5	High; protected from light	9–12 months	Suitable for heat-processed or pasteurized systems	Limited clarity; higher oxygen permeability than PET
Aluminum Pouches / Laminated Foil Packs	Outdoor, Military, High-Heat Hydration	Excellent with Na^{+} and K^{+} systems	pH 3.0–4.5	High; fully protected	12–18 months	Lightest format; durable for field applications	Seal integrity critical; not resealable; avoid carbona

							ted systems
Glass-Lined PET (Hybrid)	Premium Functional & Electrolyte Water	Excellent; prevents ion exchange	pH 2.5–4.5	Very high; near-glass performance	9–15 months	Combines clarity with inertness	Limited recyclability; costlier tooling
Bag-in-Box (BIB) Systems	Institutional / Endurance Hydration	Excellent for all ions	pH 3.0–4.5	High stability under controlled storage	6–9 months post-fill	Economical for bulk dispensing	Requires aseptic filling; not suitable for extended exposure to air
Aluminum Bottles (Powder-Coated Interior)	Ready-to-drink Isotonic or Energy Blends	Very high compatibility	Stable pH 2.5–4.0	Excellent; retains carbonation	12–18 months	Premium aesthetic; reusable; light-blocking	Higher cost; coating verification essential

Glass Jar (Pasteurized)	Recovery / Nutritional Beverages	Excellent	pH 3.5–6.0	High under pasteurization	12–18 months	Heat stable, inert	Heavier; not suitable for isotonic drinks needing portability

Stability & Formulation Insights

1. Electrolyte Interaction & Packaging Selection

- Chloride and phosphate salts are more reactive toward metals and should always be packaged in lined or coated containers (PET, glass, or epoxy-lined aluminum).
- Citrate and lactate salts are preferred for natural and clean-label systems due to their stability and low reactivity under acidic conditions.

2. Light and Oxygen Sensitivity

- PET and clear glass require UV protection when containing ascorbic acid, natural colors, or fruit extracts to prevent fading and flavor degradation.
- Amber glass, aluminum cans, and Tetra Pak offer superior protection for sensitive ingredients like vitamin C, polyphenols, or carotenoid-based colorants.

3. Thermal Processing and pH Control

- Low-pH beverages (below 3.2) are self-preserving but may corrode metal packaging without internal coatings.
- For pasteurized or aseptic products, ensure pH 3.0–4.2 to maintain both microbial safety and electrolyte stability.

4. Shelf-Life Optimization

- Maintain dissolved oxygen <0.3 ppm to preserve natural flavors and prevent oxidation of minerals.
- Use chelating agents (e.g., sodium citrate, EDTA) at trace levels to improve color and flavor retention in long-shelf-life isotonic beverages.

5. Sensory Stability

- Avoid using high-potency sweeteners alone in long-shelf-life formats, as their perception may decline over time; blend with bulk sweeteners or acids.
- Conduct accelerated storage testing (40°C/75% RH, 12 weeks) to validate flavor, color, and clarity retention across chosen packaging systems.

Common Stability Challenges and Corrective Strategies:

The following table compiles common stability and sensory challenges that arise in electrolyte systems, detailing causes, diagnostic indicators, and corrective formulation strategies.
It's designed for practical use during pilot runs, scale-up trials, and long-term shelf-life evaluations.

Table 9.2.5G: Common Stability Challenges and Corrective Strategies in Electrolyte and Hydration Beverages

Observed Issue / Instability	**Likely Root Cause**	**Diagnostic Indicators**	**Corrective Formulation Strategy**	**Preventive Best Practices**
1. Precipitation or Sediment Formation	Oversaturation of low-solubility salts (e.g., Ca^{2+}, Mg^{2+} with phosphates or sulfates); pH shift below	Visible haze or mineral sediment after 2–4 weeks	Replace insoluble salts with citrate or lactate forms; adjust pH to >3.0; reduce concentration or ionic strength	Conduct solubility screening of all salts under final pH; maintain Ca/Mg

	solubility threshold			below 0.1% w/v
2. Flavor Fading / Dullness Over Time	Oxidation of fruit flavors or sweeteners; high oxygen ingress; light exposure	Decrease in aroma, brightness, or sweetness; "flat" taste	Use ascorbic acid or natural antioxidants; add chelating agent (Na-citrate, EDTA); switch to UV-protective packaging	Nitrogen-flush headspace; maintain dissolved O_2 <0.3 ppm
3. pH Drift During Storage	Buffer imbalance; ingredient interaction (e.g., citrate vs. phosphate systems); microbial activity in low-acid formulas	pH changes >0.3 units during 3 months storage	Reinforce with dual-buffer system (citric + phosphate or citrate + lactate); revalidate preservative levels	Regular pH stability testing at 4°C, 25°C, and 40°C
4. Color Degradation / Browning	Light exposure, vitamin oxidation, reaction with trace metals (Fe, Cu)	Visible discoloratio n, especially in clear PET or glass	Include chelating agents; lower storage temperature; reduce exposure to UV	Use amber glass, cans, or multilayer PET; control trace metal contaminati on

5. Bitterness or Metallic Off-Notes on Aging	Hydrolysis of magnesium or potassium chloride; sweetener degradation	Increasing bitterness after 1–2 months	Replace chloride forms with Mg-lactate / K-citrate; rebalance acid profile	Accelerated sensory testing; maintain low ionic strength
6. Cloudiness / Haze in Clear Beverages	Poor salt solubility; pH near precipitation threshold; interaction with proteins, fibers, or pectin	Turbidity formation under temperature cycling	Replace $CaCl_2$ with Ca-lactate; pre-dissolve Mg salts; control pH >3.0	Filter final beverage to 0.45 µm; avoid incompatibl e stabilizers
7. Sweetness Reduction or Flavor Imbalance	Loss of high-intensity sweetener potency; acid–sweetener mismatch	Perceived loss of sweetness; altered flavor intensity	Use sweetener blends (e.g., stevia + erythritol); adjust acid ratio for brightness	Include sensory validation at 3–6 months under storage
8. Carbonation Loss (Sparkling Hydration)	Inadequate CO_2 sealing; high ionic concentration reduces gas retention	Flat or low-effervescen ce after 30 days	Lower osmolality (<300 mOsm/kg); optimize CO_2 pressure; switch to crown or aluminum cap	Maintain high fill-level and CO_2 retention coatings

9. Viscosity Increase / Gelation	Excess hydrocolloids; protein–electrolyte complexation; low pH	Thickening or sediment near bottom	Reduce gum concentration; avoid mixing pectin with high Ca^{2+} systems	Test hydrocolloid compatibility; maintain pH >3.5
10. Vitamin or Functional Additive Degradation	Light or oxygen exposure; metal ion catalysis	Reduced label claim potency; off-flavor from oxidation	Use microencapsulated or protected vitamin blends; add antioxidants	Store below 25°C; use opaque or barrier packaging
11. Metallic Taste from Packaging Interaction	Poor lining integrity in metal packaging; high-chloride system	Metallic or tangy off-taste after 4–8 weeks	Switch to epoxy-free internal coatings; use citrate salts to reduce reactivity	Conduct migration tests for ion leaching (Al, Fe)
12. Foam Formation or Poor Pourability	Excess carbonation; surfactant effect from flavors or electrolytes	Persistent foam layer, slow clearing	Reduce carbonation pressure; add anti-foam agents (e.g., simethicone at ppm levels)	Optimize filling conditions; degas during mixing
13. Layer Separation / Phase Instability	Inadequate emulsifier or homogenization for oil-based flavors	Visible oil ring or phase separation	Add acacia gum or modified starch emulsifier; improve homogenization	Verify emulsion stability under accelerated storage

			(3,000–5,000 psi)	
14. Sweetener Crystallization (Sucrose/Fructose)	High solids concentration; low storage temperature	White crystals or sediment at bottle base	Reduce total soluble solids (<10%); use invert syrup or maltodextrin blend	Maintain constant storage temperature (≥15°C)
15. Excess Saltiness in Final Product	High NaCl or KCl ratio; insufficient acid or sweetness balance	Harsh or briny taste; low consumer acceptability	Replace with Na-citrate / K-citrate; increase citric acid for masking	Conduct taste mapping at varied dilution ratios

Formulation & QA Insights:

- Always test ingredient compatibility at intended ionic strength and pH before finalizing the formula.
- Conduct accelerated shelf-life studies (40°C / 75% RH for 12 weeks) to simulate one year of storage.
- Validate flavor and color stability under light exposure (UV-A / UV-B) for clear-packaged products.
- When scaling up, recheck electrolyte solubility under production mixing conditions (temperature, agitation, water quality) to avoid micro-precipitates.
- Apply Good Manufacturing Practices (GMP) and water deionization to prevent trace metal contamination that accelerates instability.

Each row in the following table gives target osmolality, recommended ionic ranges (mg/L), preferred salt forms, and sensory / functional notes so you can quickly adapt blends to the product goal and manufacturing constraints.

All ionic ranges are shown as typical per 1 L finished beverage. Sodium (Na^+) estimates assume common salt forms (NaCl, Na-citrate): values are given as elemental ion mg/L for clarity.

Electrolyte Optimization:

Table 9.2.5H: Electrolyte Optimization by Beverage Function

Beverage Function	Target Osmolality	Na^+ (mg/L)	K^+ (mg/L)	Mg^{2+} (mg/L)	Ca^{2+} (mg/L)	Recommended Salt Forms	Sensory & Formulation Notes
Rehydration, Hypotonic (rapid fluid uptake)	< 270 mOsm/kg	150–300 mg	80–150 mg	5–15 mg	5–15 mg	Na-citrate / low NaCl; K-citrate; Mg-lactate; Ca-gluconate	Keep total solutes low for fast gastric emptying. Use citrate/lactate salts to soften taste. Target light, refreshing mouthfeel.
Performance, Isotonic (in-activity)	280–310 mOsm/kg	250–400 mg	120–200 mg	10–25 mg	10–25 mg	NaCl + Na-citrate blend; K-citrate; Mg-citrate or Mg-lactate; Ca-lactate	Balance immediate Na replacement with palatability. Carb level ~4–7% complements isotonic osmolality.

							Use masking acids (citric+malic).
Recovery, Hypertonic (glycogen + repair)	> 330 mOsm/kg	350–550 mg	150–300 mg	15–40 mg	15–40 mg	NaCl + Na-citrate; K-citrate; Mg-lactate; Ca-gluconate	Higher sodium supports rehydration with caloric load. Expect fuller mouthfeel; pair with protein/BCAA systems and stronger flavors to mask mineral intensity.
Functional Hybrid, Clean-label (daily wellness)	200–270 mOsm/kg	150–300 mg	80–180 mg	5–20 mg	5–20 mg	Sea-salt extract or standardized mineral water; K-citrate; Mg-lactate; Ca-lactate	Emphasize natural sources (coconut, mineral water). Use lower sweetness; prioritize citrate/lacta

							te for smooth taste and buffering.
Heat / High Sweat Loss (tropical, heavy sweaters)	280–320 mOsm/kg (can be slightly higher)	400–750 mg	150–300 mg	15–40 mg	15–40 mg	Higher NaCl (with portion replaced by Na-citrate); K-chloride + K-citrate; Mg-lactate	High Na needed to replace heavy sweat; stronger acid/flavor required to mask saltiness. Consider electrolyte concentrates or sachets.
Low-Sugar / Weight-Conscious (athleisure & wellness)	< 280 mOsm/kg	150–350 mg	100–200 mg	5–20 mg	5–20 mg	Na-citrate; K-citrate; Mg-lactate; Ca-lactate	Reduce carbs; use high-intensity sweetener blends. Keep osmolality near body to preserve absorption while lowering calories.

Clinical ORS / Medical Rehydration (e.g., WHO-inspired)	~245 mOsm/kg	~75–90 mmol Na^+ ≈ 1720–2070 mg (Note: clinical ORS uses much higher Na as mmol), follow clinical recipe	~20 mmol K^+ ≈ 780 mg	~2–5 mmol Mg^{2+} ≈ 48–120 mg	variable	Glucose + NaCl + KCl or K-citrate + trisodium citrate	Clinical ORS is a medical formulation, follow WHO/medical recipes exactly for mmol-based dosing. Not interchangeable with sports drinks.

Practical conversion & formulation notes (important):

- Elemental conversions: mg values above are elemental ion (Na^+, K^+, Mg^{2+}, Ca^{2+}) not grams of salt. When converting to salt weights, use the molar mass & ion fraction (e.g., NaCl ~39.34% Na by weight).

- Salt form selection: prefer citrate & lactate salts for consumer-facing products, they improve taste, add buffering capacity, and reduce metallic/bitterness issues common with chlorides. Use chloride forms when aggressive sodium replacement is required (e.g., extreme sweat loss), but offset sensory issues with acids/flavors.
- Magnesium & calcium: keep levels relatively low to avoid bitterness or haze. Use lactate/gluconate forms for solubility and mild flavor.
- Osmolality control: carbs (g/L) and electrolyte mg levels both contribute, always test finished osmolality rather than relying solely on individual ingredient sums.
- Sensory strategy: higher ionic loads need stronger acids, aromatics, or fruit systems to mask mineral taste. For clean-label positioning, use natural fruit acids (citric/malic) and mineral bases (coconut, mineral water).

Quick Formulator Workflows:

1. Define product purpose → pick target osmolality & carb level.
2. Select primary Na and K targets according to table range.
3. Choose salt forms: default to citrate/lactate; add chloride only if required.
4. Balance sensory: add acids (citric/malic) and select sweetener system to mask off-notes.
5. Test osmolality, pH, clarity, and taste at lab scale; iterate.
6. Run accelerated stability and sensory panels before scale-up.

9.3 Customizing formulations for sweat rate and activity type

No two athletes, or environments, are the same. The rate and composition of sweat vary widely depending on climate, exercise intensity, body size, and acclimatization level. This section guides formulators through tailoring electrolyte systems to match these diverse needs. For light or recreational activity, lower sodium and carbohydrate concentrations maintain hydration without excess calories or mineral taste. For endurance or high-heat conditions, elevated sodium and potassium levels help preserve plasma volume and prevent hyponatremia. Recovery-oriented systems may integrate magnesium and calcium for muscle relaxation and repair. Readers will learn to interpret sweat rate data and use it to inform precise electrolyte targeting, ensuring the beverage supports the body's physiological demands under specific stress conditions. By customizing the electrolyte matrix to activity type, formulators can deliver truly performance-specific hydration systems, not just generic sports drinks, but tailored solutions optimized for real-world performance.

9.3.1 Understanding Sweat Rate Variability

Sweat rate is the key driver of hydration demand and electrolyte depletion. It is influenced by body mass, metabolic heat production, ambient temperature, humidity, acclimatization, and even clothing. Typical sweat rates range from 0.5 to 2.5 L per hour, with sodium concentrations varying from 20 to 80 mmol/L. High-sweat individuals ("salty sweaters") lose more sodium and chloride relative to water, increasing their risk of hyponatremia if they consume plain water or hypotonic beverages alone.

9.3.2 Typical sweat rates and mineral losses by activity type

Hydration needs are never one-size-fits-all; they shift dramatically with the intensity, duration, and environment of physical activity. Sweat rate and mineral loss patterns provide critical insight into how the body responds under different conditions, and ultimately guide how electrolyte beverages should be formulated for specific users. This section examines the real-world variability of fluid and electrolyte loss across sports, climates, and exertion levels, translating physiological data into practical formulation targets. By understanding how athletes and consumers lose fluids and minerals in the field, developers can design beverages that meet needs with accuracy, purpose, and scientific grounding.

Table 9.3.2: Typical Sweat Rates and Electrolyte Loss by Activity Type

Activity Type	Average Sweat Rate (L/h)	Sodium Loss (mg/L)	Potassium Loss (mg/L)	Recommended Beverage Type	Formulation Notes
Endurance running / cycling	1.0–2.5	800–1800	150–250	Isotonic (6–8% carb)	Balanced Na^+/K^+ replacement; mild acidity for refreshment
Team sports (soccer, basketball)	0.8–2.0	600–1200	100–200	Isotonic (4–6% carb)	Combine quick and slow carbohydrates; flavor variety important
Strength / power training	0.5–1.0	500–800	100–150	Slightly hypotonic (3–5% carb)	Lower osmotic load; add magnesium for cramp prevention
Indoor fitness / yoga	0.3–0.8	300–600	60–120	Hypotonic (2–4% carb)	Focus on rapid absorption and light flavor; low sodium taste profile

Hot climate work / field activity	1.0–2.0	800–1600	150–250	Isotonic to slightly hypertonic	High sodium concentration, sustained electrolyte retention
Recovery / post-event	0.4–1.0	400–800	80–200	Hypertonic (8–12% carb)	Replenish glycogen and plasma sodium; may include amino acids

9.3.3 Personalizing Sodium and Carbohydrate Levels

Hydration effectiveness depends on aligning sodium and carbohydrate concentrations with the sweat profile of the target user. For instance:

- High-sodium sweaters benefit from formulations containing 700–1000 mg Na^+/L, paired with moderate carbohydrate loads (4–6%) to optimize osmolality and absorption.
- Low-sodium or light sweaters require gentler systems (300–600 mg Na^+/L) with higher fluid-to-solute ratios, often best achieved through hypotonic drinks.
- Heavy sweaters in hot conditions often require drinks with both elevated sodium and potassium to counter excessive electrolyte dilution.
- Endurance athletes need carbohydrate energy (5–8%) alongside electrolytes to delay fatigue and sustain glucose levels.

Adjusting carbohydrate type is equally important. Glucose and sucrose enable rapid uptake, while maltodextrin or cluster dextrin provide sustained energy release with minimal GI distress. The balance should maintain osmolality between 200–320 mOsm/kg, ensuring fluid absorption without gastric discomfort.

9.3.4 Environmental and Climatic Adjustments

Sweat rate increases dramatically in hot or humid conditions, and even mild dehydration (1–2% body weight loss) can impair thermoregulation and performance.
Formulation strategies should therefore consider:

- Hot, humid climates: increase sodium by 25–40%, use lower acidity and lighter flavor profiles to encourage higher fluid intake.
- Cool or dry climates: reduce sodium slightly, and use more pronounced flavor and sweetness to sustain drinking motivation.
- High altitude: prioritize hypotonic systems with moderate sodium and glucose for faster absorption and oxygen utilization efficiency.

In commercial product design, these adjustments may translate into regional SKUs (e.g., "Tropical Hydration" vs. "Mountain Recovery"), maintaining the same brand core while optimizing electrolyte ratios for climate-specific performance.

9.3.5 Activity-Specific Functional Additions

Beyond electrolytes and carbohydrates, functional co-ingredients can enhance formulation performance:

- Endurance sports: add magnesium and B-vitamins to support energy metabolism and prevent cramping.
- Power and strength training: include creatine-compatible electrolytes or branched-chain amino acids (BCAAs).
- Precision or cognitive sports (e.g., golf, esports): integrate mild nootropics (L-theanine, ginseng) for focus and calm alertness.
- Heat-stress applications: consider osmoprotectants such as betaine or glycerol for fluid retention.

9.3.6 Formulation Insight

- Sweat rate customization transforms hydration beverages from generic to personalized performance tools.

- Maintain Na^+:K^+ ratio between 3:1 and 5:1 for most isotonic systems.
- Align osmolality with activity type: hypotonic (200–250 mOsm/kg) for rapid rehydration, isotonic (270–310 mOsm/kg) for performance, and hypertonic (>330 mOsm/kg) for recovery.
- Always validate formulations through field testing under realistic activity and climate conditions.

9.4 Color, Clarity, and Aroma as Brand Identifiers

Color, clarity, and aroma are among the first sensory cues consumers use to judge a beverage's quality, purity, and function. In functional hydration products, these visual and olfactory attributes must not only reinforce the brand identity but also signal freshness, performance, and trustworthiness. A clear, bright, and naturally scented product communicates hydration and vitality, while opacity, cloudiness, or off-odor can imply instability or artificiality. This section explains how to engineer color, clarity, and aroma systems that support both sensory experience and product stability, transforming visual and aromatic design into a tool for brand differentiation.

9.4.1 Color as a Functional and Psychological Signal

Color selection in hydration beverages extends beyond aesthetics, it conveys performance cues and emotional triggers:

- Light, cool hues (clear, blue, green) suggest freshness, thirst-quenching ability, and low-calorie purity.
- Warm, vibrant tones (orange, red, yellow) evoke energy, endurance, and flavor intensity, ideal for isotonic or recovery products.
- Natural tints (pale amber, soft pink) enhance clean-label positioning and align with fruit- or plant-based claims.

Colorants must remain stable under light, pH, and thermal processing. Formulators commonly use:

- Natural pigments: beta-carotene, anthocyanins, turmeric, or fruit/vegetable extracts for clean-label appeal.

- Nature-identical or synthetic colors: when clarity, cost, or stability outweighs labeling constraints.

Key parameters to control:

- pH range stability: e.g., anthocyanins shift hue with acidity; best used below pH 4.
- Light resistance: avoid chlorophyll-based pigments in transparent packaging unless UV-blocking bottles are used.
- Concentration: target 0.005–0.02 % w/v for clear hydration products to maintain visual transparency.

9.4.2 Clarity and Visual Purity

Clarity equals consumer trust. Turbidity, haze, or sedimentation suggests instability, even if harmless, and can reduce perceived quality. Achieving optical clarity requires control over dissolved solids, ionic strength, and ingredient compatibility.

Strategies for maintaining clarity:

1. Use highly soluble electrolyte salts (e.g., sodium citrate, potassium chloride) and avoid carbonate or sulfate forms prone to precipitation.
2. Maintain pH between 3.2 and 4.2, balancing acid stability and microbial safety.
3. Select transparent sweeteners and flavor carriers (liquid glucose, glycerol esters) that fully dissolve.
4. Implement fine filtration (0.2–0.45 µm) post-mixing and use de-aeration to remove microbubbles.
5. Avoid incompatible combinations: calcium + phosphate or protein systems cause turbidity. Use lactate or gluconate forms instead.

Formulation Note:

Adding small amounts of chelating agents (e.g., 20–40 ppm sodium citrate) can bind free metal ions that would otherwise cause haze or color darkening, without affecting taste.

9.4.3 Aroma: The Emotional Gateway to Hydration

Aroma defines the beverage's identity before it even touches the lips. It shapes perceived freshness, flavor authenticity, and brand memorability. Hydration drinks typically emphasize citrus, berry, tropical, or herbal top notes associated with refreshment and vitality.

Aroma Design Principles:

- Top notes: light, volatile esters (lemon, lime, grapefruit, cucumber) create instant freshness.
- Middle notes: fruity or floral elements (mango, peach, jasmine) give fullness and differentiation.
- Base notes: subtle grounding tones (mint, ginger, vanilla) provide stability and after-feel continuity.
- Functional pairings: electrolytes pair best with bright citrus and tart fruit aromas; protein or recovery blends benefit from creamy or cocoa profiles.

Technical Considerations:

- Encapsulation (e.g., spray-dried or micro-encapsulated flavor oils) minimizes aroma loss during hot-fill or pasteurization.
- pH and temperature: volatile esters degrade rapidly above 50 °C; cold-fill systems preserve delicate notes.
- Solvent selection: use natural carriers such as ethanol (<0.1 %) or propylene glycol for consistent dispersion.

9.4.4 Color, Clarity, and Aroma Design Guide

Color, clarity, and aroma form the sensory signature of a hydration beverage, shaping how consumers perceive function, purity, and performance before the first sip. This guide breaks down how to intentionally design these attributes, not as decorative elements, but as strategic formulation tools. From selecting color systems that signal functionality, to engineering optical clarity that communicates quality, to building aroma profiles that enhance flavor expectation and emotional appeal, this section provides a practical framework for turning sensory choices into a cohesive, stable, and market-ready identity.

Table 9.4.4: Color, Clarity, and Aroma Design Guide - Sensory–Stability Integration for Hydration Beverages

Product Type	Visual Goal	Preferred Color Family	Typical Clarity Target (NTU)	Aroma Direction	Formulation Notes
Hypotonic (Rehydration)	Light, refreshing	Clear / Pale blue / Citrus yellow	< 2 NTU	Lemon–lime, cucumber, mint	Keep color dilution high; use low-acid freshness cues
Isotonic (Performance)	Energetic, bold	Orange, red, bright yellow	2–5 NTU	Citrus, berry, tropical	Slight turbidity acceptable; use esters with high heat stability
Hypertonic (Recovery)	Nourishing, dense	Amber, ruby, creamy beige	5–10 NTU	Vanilla, cocoa, banana	Protein or carb base may reduce clarity; manage with emulsifiers
Functional (Clean-Label Hybrid)	Natural, minimalistic	Soft green, pale pink, transparent	< 3 NTU	Herbal, fruit-infused	Natural pigments only; avoid artificial aromas

Sparkling Electrolyte Water	Crystal-clear brilliance	None / tintless	< 1 NTU	Citrus zest, neutral fruit	Use colorless flavor extracts; ensure CO_2 retention stability

9.4.5 Integrating Visual and Aromatic Identity into Branding

Color and aroma consistency are as important as formula reproducibility.

- Color: serves as an instant brand cue, consumers associate specific hues with performance level (e.g., orange = energy, blue = hydration, green = natural).
- Aroma: builds brand memory, consistent scent signatures increase recognition and trust.
- Clarity: communicates technological sophistication and purity, reinforcing quality perception.

For startups and established formulators alike, designing an intentional sensory identity transforms hydration products from functional commodities into memorable experiences. Harmonizing color, clarity, and aroma with flavor and packaging design ensures the beverage's first impression reflects its physiological performance and brand promise.

Formulation Insight

- Keep pigment and aroma systems modular, allow easy reformulation for flavor extensions without altering electrolyte base.
- Conduct accelerated light and heat testing (40 °C / 10 days) to confirm pigment and aroma stability.
- Match visual language to functional promise: clarity = purity, brightness = energy, aroma = authenticity.

10. Texture and Stability Management

A hydration beverage's technical success extends far beyond its electrolyte balance or flavor, it must remain physically stable, visually appealing, and texturally consistent throughout its shelf life. Consumers expect clarity in isotonic drinks, smoothness in recovery shakes, and crisp refreshment in functional waters. To meet these expectations, formulators must manage colloidal stability, prevent sedimentation, and control interactions between minerals, carbohydrates, and proteins that can destabilize the system.

This chapter provides practical tools for building stable, high-performance hydration beverages that maintain their intended texture and functionality from production to consumption.

10.1 Preventing Sedimentation and Phase Separation

Sedimentation and phase separation are two of the most common stability challenges in electrolyte and nutrition beverages. They arise when suspended particles, minerals, or emulsified oils lose colloidal balance and settle or separate during storage. Even minor visual defects can undermine consumer confidence and signal poor formulation control.

10.1.1 Causes of Sedimentation

Sedimentation occurs when insoluble or semi-soluble solids (such as mineral salts, plant extracts, or protein particles) exceed the solution's capacity to keep them suspended. Key causes include:

- High ionic strength from excessive electrolyte loading, reducing electrostatic repulsion between particles.
- Inadequate viscosity or lack of stabilizers to slow gravitational settling.
- pH shifts leading to reduced solubility of salts or proteins (e.g., calcium phosphate precipitation).
- Temperature fluctuations during storage that alter solubility equilibrium.

- Mechanical stress (transport vibration) accelerating particle settling.

10.1.2 Causes of Phase Separation

Phase separation typically affects emulsified or multi-component systems, especially those containing flavor oils, protein–carbohydrate matrices, or lipid-based additives. It manifests as visible layering, oil rings, or cloud collapse. Contributing factors include:

- Poor emulsifier or stabilizer selection for oil–water systems.
- Mismatch of density between dispersed and continuous phases.
- Inadequate homogenization pressure or particle size distribution.
- Electrolyte-induced flocculation (high sodium or divalent ions can destabilize emulsions).

10.1.3 Stabilization Strategies

Stabilizing a hydration beverage requires understanding the forces that pull a formulation apart and the tools that hold it together. Stabilization strategies offer the practical mechanisms for controlling sedimentation, preventing phase separation, and maintaining visual and textural integrity throughout shelf life. This section outlines the core approaches, mechanical, chemical, and formulation-based, that enable developers to create beverages that remain uniform, appealing, and reliable under real-world storage and distribution conditions.

1. Optimize Solubility and Ionic Strength
    - Choose high-solubility salts (e.g., sodium citrate, potassium lactate) over less soluble ones like magnesium sulfate.
    - Keep total dissolved solids (TDS) below 1.2% for clear hypotonic drinks and 2–3% for isotonic systems to maintain clarity.
    - Monitor pH window (3.2–4.2) for most organic acid systems, outside this range, precipitation risk increases.
2. Enhance Suspension Viscosity

- Use low-level hydrocolloids (e.g., xanthan gum 0.02–0.05%, gellan gum 0.02–0.04%, or pectin 0.05–0.1%) to increase apparent viscosity and suspend particulates.
- For clear systems, select transparent-grade gums that provide stability without haze.
- For recovery or protein-rich beverages, combine guar gum or carrageenan with dairy or plant proteins to stabilize the matrix.

3. Emulsion Stabilization (Oil-Containing Systems)
    - Use emulsifiers like polysorbate 80, sucrose esters, or natural alternatives (quillaja, gum arabic) to maintain droplet uniformity.
    - Apply homogenization pressures between 150–250 bar for stable, fine dispersions in isotonic bases.
    - Match oil phase density with aqueous phase using propylene glycol or glycerol as balancing agents.
4. Thermal and Storage Stability
    - Conduct accelerated stability tests (40 °C for 14 days) and vibration simulations to evaluate separation risk.
    - Incorporate buffer systems (citric + malic acid) to resist pH drift during storage.
    - For high-protein formulations, maintain low-ionic environments and cold-fill processes to prevent protein coagulation.

10.1.4 Instability Problems and Corrective Actions

Even well-designed formulations can develop instability during production, storage, or distribution, and understanding these failure modes is essential for maintaining product quality. Instability problems and corrective actions examine the most common issues, haze formation, sedimentation, flavor oil separation, viscosity drift, and mineral precipitation, and connect each to its underlying cause. This section provides a practical troubleshooting framework, giving formulators clear diagnostic cues and targeted interventions to restore stability and prevent recurrence in future batches.

Table 10.1.4: Common Instability Problems in Hydration Beverages and Corrective Formulation Strategies

Problem	Likely Cause	Recommended Correction
Visible sediment (minerals or colorants)	Poor solubility, high TDS	Switch to citrate/lactate salts; adjust pH to 3.5–4.0
Hazy appearance	Electrolyte–acid interaction	Add chelating agents (citric acid, EDTA); fine filtration
Oil ring or cloud collapse	Insufficient emulsifier or low homogenization	Increase emulsifier dosage; re-homogenize at higher pressure
Protein precipitation	pH below isoelectric point; divalent ions	Use calcium lactate instead of chloride; adjust pH upward by 0.2–0.4
Layer separation during storage	Density mismatch or temperature cycling	Add light gums; improve thermal storage control
Gas release or bubbles	Carbonate reaction or microbial activity	Control CO_2 levels; ensure aseptic filling and preservative system

10.1.5 Formulation Insight

- Balance total ionic strength early, overly concentrated electrolyte systems are the leading cause of haze and sediment.
- Always test stability under realistic shipping conditions (temperature, vibration, UV exposure).
- For "clean-label" products, natural gums (gellan, pectin, guar) can replace synthetics if hydration and shear mixing are carefully optimized.

- Use cold water dispersion techniques for hydrocolloids to avoid clumping and ensure reproducible texture.

10.2 Stabilizers, Emulsifiers, and Gums

Stabilizers, emulsifiers, and gums form the hidden architecture of every successful hydration beverage. Though typically less than one percent of the formulation, these ingredients govern clarity, viscosity, mouthfeel, and long-term product integrity. Selecting the right combination requires understanding not just their individual functions, but how they interact with electrolytes, acids, sweeteners, and active nutrients. This section outlines the key classes of stabilizing agents used in electrolyte, isotonic, and recovery systems, and provides selection criteria based on desired sensory and functional outcomes.

10.2.1 Functional Roles in Beverage Systems

Stabilizers, emulsifiers, and gums play far more than a supporting role in functional beverages, they shape the very structure, texture, and reliability of the final product. Understanding their functional roles allows formulators to engineer drinks that remain visually uniform, resist separation, and deliver a consistent sensory experience from the first pour to the last. This section breaks down how these ingredients influence viscosity, suspension, emulsion stability, and flavor delivery, forming the technical foundation for building stable, high-performance hydration systems.

1. Stabilizers

Prevent separation of dispersed particles or phases by increasing viscosity or creating a weak gel network. They control the mobility of insoluble solids, proteins, and flavor oil droplets.

2. Emulsifiers

Enable stable dispersion of oil-based flavor or nutrient components within the aqueous phase by reducing interfacial tension. They ensure uniform distribution and prevent oil creaming or ring formation.

3. Gums / Hydrocolloids

Modify texture, mouthfeel, and body. They create suspension stability, adjust perceived thickness, and influence the release of flavor and sweetness.

Each of these components must be chosen carefully to maintain clarity, processability, and sensory harmony without altering osmolality or hydration efficiency.

10.2.2 Selection Criteria for Hydration Beverages

Choosing the right stabilizers and emulsifiers for a hydration beverage requires more than matching ingredients to function, it demands alignment with the product's sensory goals, processing conditions, and shelf-life expectations. Selection criteria for hydration beverages outlines the key considerations that guide this choice, from electrolyte compatibility and desired mouthfeel to clarity requirements, pH tolerance, and clean-label constraints. This section provides formulators with a clear decision-making framework to ensure each system supports both performance and consumer expectations.

Table 10.2.2: Selection Criteria for Hydration Beverages

Beverage Type	Target Texture	Recommended System	Notes
Hypotonic (Rehydration)	Light, crisp, water-like	Minimal stabilizer; <0.02% xanthan or gellan gum	Maintain low viscosity (<2 mPa·s) for rapid absorption
Isotonic (Performance)	Smooth, slightly full-bodied	Gum arabic + gellan gum (0.02–0.05%)	Improves mouthfeel and flavor retention
Hypertonic (Recovery)	Rich, nourishing, protein-compatible	Carrageenan + guar + lecithin or mono/diglycerides	Prevents phase separation in protein–carb matrices

Functional (Clean-Label Hybrid)	Natural, slightly structured	Pectin or acacia gum (0.05–0.1%)	Supports label claims; plant-derived hydrocolloids
Sparkling or Effervescent	Crisp, low-viscosity	None or <0.01% stabilizer	Avoid foaming; use CO_2-stable emulsifiers only

10.2.3 Common Stabilizers and Their Characteristics

A successful stabilization system begins with knowing the strengths and limitations of the ingredients available. Common stabilizers and their characteristics provide a clear, practical overview of the hydrocolloids, gums, and emulsifying agents most frequently used in hydration beverages, detailing how each behaves in solution, interacts with electrolytes, and influences texture, clarity, and stability. This section equips formulators with the foundational knowledge needed to select and combine stabilizers with precision and confidence.

Table 10.2.3: Stabilizers and Hydrocolloids for Beverage Applications

Ingredient	Source / Type	Functional Role	Optimal Use Level	Compatibility	Notes
Gellan Gum (Low Acyl)	Fermentation-derived polysaccharide	Forms fine colloidal network; suspends minerals, maintains clarity	0.02–0.04%	Excellent with acids and electrolytes	Preferred for clear isotonic beverages
Xanthan Gum	Bacterial polysaccharide	Viscosity builder, stabilizes	0.02–0.1%	Good pH and	Use low levels to avoid

		emulsions		temperature tolerance	slimy texture
Pectin (High Methoxyl)	Fruit-derived polysaccharide	Improves mouthfeel; provides smooth body	0.05–0.2%	Sensitive to high electrolytes	Ideal for natural fruit-based isotonic or recovery beverages
Guar Gum	Legume-derived galactomannan	Adds body and smoothness	0.05–0.3%	Limited acid stability	Use in protein or recovery beverages; not ideal for clear drinks
Carrageenan (κ-type)	Seaweed extract	Stabilizes milk/protein systems	0.02–0.1%	Incompatible with high acidity	For recovery shakes or protein-rich drinks
Gum Arabic (Acacia)	Natural tree exudate	Emulsifier–	0.2–0.5%	Good flavor oil	Supports clean-label

		stabilizer hybrid		compatibility	emulsions
CMC (Carboxymethylcellulose)	Cellulose derivative	Controls viscosity; prevents phase separation	0.1–0.4%	Stable in wide pH	Common in isotonic and recovery blends
Locust Bean Gum (LBG)	Carob seed polysaccharide	Enhances mouthfeel and prevents whey-off	0.05–0.2%	Synergistic with carrageenan	Best in creamy or high-carb beverages

10.3 Emulsifier Systems and Flavor Oil Stability

Electrolyte beverages containing natural flavors, vitamins, or lipid-soluble nutrients require effective emulsifiers to ensure uniform distribution and transparency. Common systems include:

10.3.1 Synthetic / Conventional Emulsifiers

Synthetic and conventional emulsifiers provide the most dependable foundation for stabilizing flavor oils, especially in electrolyte-rich or low-pH hydration systems. Their well-characterized behavior at the oil-water interface, coupled with strong tolerance to processing stress, makes them essential tools for achieving clarity, uniformity, and long-term stability. By understanding how these emulsifiers operate and where they excel, formulators can design emulsion systems that perform consistently from production to shelf.

- Polysorbate 80 (0.05–0.1%): high-performance, neutral taste; excellent for citrus oil stabilization.

- Sucrose esters (0.05–0.2%): smooth mouthfeel, works well in low-fat, high-acid beverages.

10.3.2 Natural / Clean-Label Emulsifiers

Natural and clean-label emulsifiers bring a different design philosophy to hydration beverages, emphasizing transparency, minimal processing, and consumer trust while still delivering functional stability. These systems rely on plant-based, fermentation-derived, or naturally occurring structures to suspend flavor oils and maintain uniformity, yet each comes with its own performance nuances, processing sensitivities, and compatibility limits. Understanding their behavior allows formulators to balance clean-label expectations with the technical demands of clarity, stability, and sensory consistency.

- Gum arabic (0.3–0.5%): plant-based, mild taste, supports flavor oil dispersion.
- Quillaja saponaria extract (0.02–0.04%): natural foaming/emulsifying agent; good for sparkling waters.
- Lecithin (sunflower or soy) (0.1–0.3%): ideal for recovery and protein drinks.

10.3.3 Optimization Tips

Optimizing an emulsion system requires attention to the fine details that determine long-term stability, from droplet size control and homogenization parameters to electrolyte tolerance and storage resilience. Effective optimization turns a functional emulsifier into a robust, shelf-stable solution by aligning ingredient behavior with processing conditions and sensory goals. These practical tips help formulators refine performance, troubleshoot weaknesses, and build emulsions that remain clear, consistent, and reliable throughout the product's lifecycle.

- Maintain oil droplet size <1 μm for clarity and stability.
- Use high-shear or ultrasonic homogenization during mixing.
- Adjust pH >3.0 for gum arabic stability; below this, degradation can occur.
- For transparent beverages, avoid excess emulsifier which may cause opalescence.

10.4 Synergy Between Gums and Electrolytes

Gums and electrolytes interact in ways that can either strengthen a beverage's structure or destabilize it, depending on how they are balanced. Their synergy defines key attributes such as viscosity, suspension efficiency, mouthfeel, and overall stability, making this relationship central to the design of functional hydration systems. Understanding how different gums respond to ionic strength, pH, and mineral composition allows formulators to harness these interactions deliberately, creating beverages that remain smooth, uniform, and resilient under real-world conditions.

Electrolyte systems, especially those high in sodium or divalent cations (Ca^{2+}, Mg^{2+}), can influence gum hydration and viscosity. For instance:

- Calcium interacts strongly with pectin and carrageenan, creating gels or precipitates if levels exceed 100–150 ppm free Ca^{2+}.
- Magnesium ions may thin xanthan gum dispersions by interfering with polymer hydration.
  To mitigate these effects:
- Add electrolytes after full gum hydration, ideally under continuous agitation.
- Sequence mixing: water → gum dispersion → sweeteners → acids → electrolytes → flavors.
- For multivalent systems, use chelating buffers (sodium citrate 0.1%) to sequester free metal ions.

10.4.1 Compatibility of Gums with Electrolytes

Compatibility between gums and electrolytes is a defining factor in achieving stable texture and visual uniformity in hydration beverages. Each gum responds differently to ionic strength, divalent minerals, and pH shifts, which can enhance viscosity, thin the system, or trigger unwanted gelation and precipitation. Understanding these compatibility patterns enables formulators to choose the right hydrocolloids for mineral-rich environments and engineer systems that remain consistent, stable, and sensory-balanced throughout shelf life.

Table 10.4.1: Compatibility of Gums with Electrolytes

Gum Type	Na^+ Stability	K^+ Stability	Ca^{2+} / Mg^{2+} Stability	pH Range	Optical Clarity
Gellan Gum	Excellent	Excellent	Good	3.0–7.0	High
Xanthan	Excellent	Good	Moderate	2.5–8.0	Moderate
Pectin	Moderate	Moderate	Poor	2.8–4.2	High
Carrageenan	Good	Good	Sensitive to Ca^{2+}	5.0–8.0	Opaque
Gum Arabic	Excellent	Excellent	Excellent	3.0–6.5	Very High

10.4.2 Formulation Insight

Effective formulation hinges on recognizing how gums behave in the presence of electrolytes and using that knowledge to shape stability, texture, and sensory flow. Key insights emerge when examining how ionic strength influences viscosity, how mineral type affects hydration and dispersion, and how small adjustments in concentration can prevent phase shifts or gel formation. By applying these principles strategically, formulators can build hydration systems that maintain clarity, uniformity, and functional performance under a wide range of processing and storage conditions.

- For clear isotonic drinks, gellan gum remains the gold standard due to its dual ability to stabilize minerals and maintain transparency.
- Avoid overuse of gums; excessive viscosity suppresses perceived refreshment and may increase osmolality.
- Always hydrate gums in preheated water (70–80°C) before combining with acids or salts for full activation.
- Conduct centrifuge testing (3000 rpm, 15 min) during pilot trials to predict long-term phase stability.

- Match stabilizer selection to the desired sensory signature: crisp and light for hypotonic, smooth for isotonic, and creamy or nourishing for recovery beverages.

10.5 Shelf-Life Extension and Clarity Optimization

Shelf life defines not only how long a beverage remains microbiologically safe, but also how long it retains its intended sensory and functional properties, color, flavor, clarity, and mouthfeel. In electrolyte and nutrition beverages, the challenge lies in extending stability without compromising freshness or "clean-label" appeal. Because these products are often rich in salts, carbohydrates, or bioactive compounds, maintaining clarity and preventing flavor or color drift over time requires a careful balance of formulation chemistry, process control, and packaging selection.

10.5.1 Key Stability Challenges in Hydration Beverages

Hydration beverages face a unique set of stability challenges driven by their high-water activity, electrolyte load, acidic pH, and often delicate flavor or color systems. These factors create conditions where haze, sedimentation, phase separation, flavor degradation, and nutrient loss can easily develop if not carefully managed. Understanding these core instability drivers provides the foundation for designing formulations that remain clear, uniform, and sensorially consistent throughout their intended shelf life.

1. Precipitation and Haze Formation
    - Caused by insoluble salts (e.g., Ca^{2+} and phosphate interaction), pH drift, or incompatible flavor systems.
    - Leads to reduced visual appeal and perceived instability.
2. Color Degradation
    - Natural pigments like anthocyanins or beta-carotene are sensitive to light, oxygen, and temperature.
    - Fading or browning often occurs within 3–6 months if not properly protected.

3. Flavor Oxidation
    - Citrus and tropical notes are particularly susceptible to oxidative loss, producing "stale" or "cardboard" flavors.
    - Trace metals, oxygen ingress, and UV exposure accelerate these reactions.
4. Microbial Growth (in Low-Acid Systems)
    - pH > 4.5 and inadequate pasteurization or preservatives allow yeast or mold proliferation, especially in protein-based drinks.
5. Viscosity Drift
    - Hydrocolloids may lose water-binding over time due to ionic stress or thermal cycles, leading to thinning or phase separation.

10.5.2 Clarity Optimization Principles

A clear, bright beverage signals purity and quality, but clarity must be chemically engineered. Maintaining optical stability depends on controlling solubility, ionic balance, and micro-particulate dispersion.

Core strategies:

- Maintain Turbidity < 2 NTU for transparent isotonic and hypotonic drinks.
- Keep pH within 3.3–4.0 for acidified electrolyte systems to ensure both clarity and microbial safety.
- Use chelating buffers (citric, malic, or tartaric acid salts) to bind trace metals that cause color shift or haze.
- Prefer citrate or gluconate forms of electrolytes over carbonates or sulfates, which often crystallize on storage.
- Conduct cold stability tests (4°C for 7 days) and heat tests (40°C for 10 days) to predict shelf haze formation.

10.5.3 Instability Modes and Prevention Techniques

Instability in hydration beverages can arise from multiple mechanisms, whether driven by physical separation, chemical degradation, or interactions among electrolytes, flavors, and stabilizers. Recognizing these distinct modes is essential for preventing haze formation, sedimentation, emulsion breakdown, color fading, and textural drift over time. By understanding the root causes and applying targeted prevention techniques, formulators can build systems that maintain clarity, uniformity, and sensory integrity across processing, storage, and real-world handling.

Table 10.5.3: Common Instability Modes and Prevention Techniques in Functional Hydration Beverages

Instability Type	Root Cause	Prevention Strategy	Recommended Additives
Mineral haze / sediment	Insoluble salts (e.g., Ca, Mg, phosphate)	Use lactate or citrate salts; adjust pH to 3.5–4.0	Sodium citrate (0.05–0.1%), sequestering agents
Color fading	UV or oxidative degradation of natural pigments	Use opaque or amber packaging; add antioxidants	Ascorbic acid (100–200 ppm), rosemary extract
Flavor oxidation	Oxygen ingress, metal catalysis	Deaerate mix; use nitrogen headspace flushing	Ascorbate, EDTA (clean-label alternatives: green tea polyphenols)
Loss of aroma	Volatile ester evaporation	Use microencapsulated flavor oils; cold-fill processing	Spray-dried citrus oil, gum arabic
Microbial spoilage	Inadequate acidity or heat treatment	Maintain pH < 4.2; apply pasteurization or natural preservatives	Potassium sorbate, natamycin, or rosemary extract
Viscosity loss	Hydrocolloid breakdown under ionic stress	Use gellan or CMC instead of guar; monitor electrolyte ratio	Gellan gum (0.03%), sodium citrate

10.5.4 Formulation Tools for Extended Shelf Life

Extending shelf life in hydration beverages depends on selecting the right combination of stabilizers, processing methods, and protective ingredients that can defend against physical, chemical, and sensory degradation. Formulation tools such as optimized emulsifier systems, chelators, antioxidants, clarity enhancers, and controlled pH environments work together to preserve uniformity and flavor integrity over time. By applying these tools strategically, developers can engineer beverages that remain stable, appealing, and reliable throughout their full distribution and storage cycle.

1. pH Control and Buffering
   - A stable pH prevents both microbial growth and precipitation.
   - Buffer systems (citric–malic or phosphate–citrate) maintain acid balance even with temperature shifts.
   - Use titration testing to establish buffer capacity across expected storage temperatures.
2. Oxygen and Light Management
   - Oxygen accelerates pigment and flavor oxidation.
   - Use nitrogen sparging before filling and light-blocking PET or glass bottles.
   - For clear packaging, integrate UV-absorbing masterbatches or protective sleeve labels.
3. Antioxidant Systems
   - Combine ascorbic acid (vitamin C) with tocopherols or plant polyphenols for dual-phase protection.
   - Synergistic blends (e.g., ascorbate + green tea extract) extend flavor and color stability by up to 50%.
   - Maintain antioxidant levels below sensory thresholds to avoid bitterness.
4. Preservative Optimization

- Use preservatives compatible with target pH and clean-label goals.
- For natural positioning, leverage fermented sugar acids (lactic, gluconic, or levulinic acid) as natural antimicrobials.
- For extended ambient stability (>6 months), consider hot-fill pasteurization (85–90°C, 30–60 sec) or cold-fill aseptic processing.

5. Packaging Engineering
    - PET bottles: lightweight, suitable for isotonic drinks; must include UV barrier for natural colors.
    - Aluminum cans: best oxygen barrier, ideal for sparkling electrolyte beverages.
    - Glass: premium clarity and inertness; use nitrogen headspace to protect aromas.
    - Multi-layer laminates (Tetra Pak): superior light and oxygen barrier for clean-label or plant-based recovery beverages.

10.5.5 Shelf-Life Validation Protocol

Validating shelf life requires more than observing whether a beverage simply "holds together" over time, it demands a structured protocol that tests how the formulation responds to heat, light, oxygen exposure, mechanical stress, and real-world storage conditions. A disciplined validation workflow reveals the early signs of instability, quantifies degradation rates, and confirms whether the product can maintain its intended sensory and functional qualities from production to expiration. By following a rigorous protocol, formulators gain the data needed to ensure reliability, regulatory compliance, and consumer trust.

Table 10.5.5: Shelf-Life Validation Protocol

Test Type	Conditions	Duration	Acceptance Criteria
Accelerated Shelf Life	40°C ± 2°C	14 days	No haze, flavor drift <10%, pH variation <0.1

Real-Time Stability	25°C / 60% RH	6–12 months	Stable color, clarity, and sensory rating ≥8/10
Cold Stability	4°C storage	7 days	No crystallization or precipitation
Transport Simulation	35°C + vibration	48 hours	No phase separation or foaming
UV Light Exposure	1,000 lux for 72 hours	–	<5% color loss ($\Delta E < 2.0$)

10.5.6 Formulation Insight

Actionable insight emerges when shelf-life data is translated back into formulation decisions that strengthen stability and sensory resilience. Understanding which variables, pH, mineral load, emulsifier choice, antioxidant systems, or processing steps, most strongly influence long-term performance allows developers to refine the beverage architecture with precision. These insights help transform raw stability findings into practical adjustments that enhance clarity, flavor retention, and overall product durability across real-world conditions.

- Clarity and shelf stability are interconnected, improving ionic compatibility often enhances optical quality.
- Use low-oxygen water (<2 mg/L dissolved O_2) and maintain closed-loop mixing during production.
- Test product clarity at both cold (4°C) and ambient (25°C) storage, since some salts precipitate at lower temperatures.
- Integrate sensory evaluation checkpoints (color, taste, mouthfeel) into stability studies to capture holistic quality shifts.
- For clean-label beverages, rely on natural antioxidants and multi-barrier packaging rather than synthetic preservatives.

11. Sweetness Architecture and Caloric Balancing

Sweetness is more than taste; it is a structural and emotional component of hydration beverage design. In isotonic and recovery drinks, it provides energy and enhances palatability; in hypotonic and electrolyte waters, it must remain subtle and refreshing. Beyond flavor, the choice and balance of sweeteners determine osmotic pressure, absorption rate, and even how electrolytes are perceived on the palate.

This chapter examines the architecture of sweetness, how to combine natural and functional sweeteners to achieve both sensory appeal and physiological compatibility, while managing calories and regulatory constraints.

11.1 Understanding Sweetness Dynamics in Hydration Beverages

The perception of sweetness is influenced by intensity, onset, duration, and aftertaste. Unlike table sugar, alternative sweeteners often have different time–intensity curves, so their combination defines the "sweetness shape" of a beverage.

Key aspects of sweetness dynamics:

- Onset speed: Determines initial taste recognition (important for "refreshment" perception).
- Peak intensity: Maximum sweetness perceived after swallowing.
- Persistence: How long sweetness lingers; excessive persistence can clash with electrolytes.
- Thermal and pH stability: Crucial for hot-fill or acidified systems.

Sweetness must harmonize with the product's hydration goal:

- Hypotonic drinks: low sweetness (1.5–2.5 °Brix equivalent) for rapid drinking and thirst quenching.

- Isotonic drinks: moderate sweetness (4–6 °Brix equivalent) for energy and palatability balance.
- Hypertonic / recovery drinks: higher sweetness (6–10 °Brix equivalent) to support glycogen replenishment and perceived nourishment.

11.1.1 Sweetener Profiles

Sweetener profiles define far more than sweetness level, they shape flavor release, aftertaste, mouthfeel, and even hydration performance. Each sweetener carries a unique sensory curve, interaction pattern, and functional behavior within electrolyte-rich, acidic beverage systems. Understanding these profiles allows formulators to select sweetening systems that align with the product's purpose, delivering clean sweetness, balanced perception, and stability without compromising clarity or performance.

Table 11.1.1: Sweetener Profiles and Technical Characteristics

Sweetener	Sweetness vs. Sucrose (×)	Caloric Value (kcal/g)	Key Characteristics	Typical Use Level	Notes
Sucrose	1.0	4.0	Clean taste, natural benchmark	3–10%	Increases osmolality; use carefully in isotonic blends
Glucose (Dextrose)	0.7	3.8	Rapid energy source, mild sweetness	3–8%	Supports fast absorption in endurance formulas

Fructose	1.2–1.8	4.0	High sweetness, low glycemic index	1–6%	Use sparingly; excessive levels may delay gastric emptying
Maltodextrin	0.1–0.2	3.8	Adds body, low sweetness	2–10%	Ideal for recovery drinks and neutral taste
Stevia (Reb-A or Reb-M)	200–300×	0	Natural high-intensity sweetener	0.01–0.05%	May impart bitterness; best used in blends
Sucralose	600×	0	Heat- and pH-stable	0.005–0.02%	Clean taste; supports low-calorie isotonic systems
Acesulfame-K	200×	0	Fast onset, synergistic with sucralose	0.01–0.03%	Slight metallic aftertaste if overused
Erythritol	0.7	0.2	Natural sugar alcohol, cooling effect	1–3%	Useful in sugar-free hydration waters
Allulose	0.7	0.2	Rare sugar, near-zero calorie	2–5%	Adds body; similar

					mouthfeel to sucrose
Trehalose	0.5	4.0	Stable disaccharide, mild sweetness	2–8%	Stabilizes proteins and aromas; good for recovery beverages

11.1.2 Sweetness Synergy and Blending

Sweetness perception can be optimized through synergistic blending, where two or more sweeteners create a smoother, more natural taste profile while masking off-notes.

Typical pairings include:

- Sucralose + Acesulfame-K: Delivers fast onset and clean finish, ideal for isotonic beverages.
- Stevia + Erythritol: Provides a natural, low-calorie sweetness with balanced mouthfeel and reduced bitterness.
- Fructose + Glucose: Mimics sucrose's sweetness curve, improving palatability in sports formulations.
- Maltodextrin + Sucralose: Adds body and sweetness with minimal calorie impact.

Formulation Tip: Blending 0.012% sucralose + 0.015% acesulfame-K yields a sweetness equivalent to ~6% sucrose, with minimal osmotic impact, perfect for zero-calorie isotonic beverages.

11.1.3 Physiological Impact and Osmolality Balance

Sweeteners influence fluid absorption through their contribution to osmolality, the concentration of solutes that determine water transport across intestinal membranes.

- Low osmolality (<250 mOsm/kg): Promotes rapid hydration (ideal for hypotonic drinks).
- Isotonic range (270–320 mOsm/kg): Balances absorption and energy release.
- High osmolality (>330 mOsm/kg): Slows absorption; suited for recovery or post-workout drinks.

To maintain absorption efficiency, total carbohydrate content should generally not exceed:

- 3–4% w/w for rehydration formulations
- 6–8% w/w for performance isotonic drinks
- 10–12% w/w for recovery blends

11.1.4 Recommended Sweetness and Caloric Targets

Recommended sweetness and caloric targets provide the quantitative backbone for designing hydration beverages that meet both functional needs and consumer expectations. Striking the right balance influences not only taste but also osmolality, absorption rate, energy delivery, and overall positioning within the hydration spectrum. By grounding formulation decisions in evidence-based sweetness ranges and caloric thresholds, developers can create drinks that are palatable, physiologically effective, and aligned with modern nutritional standards.

Table 11.1.4: Recommended Sweetness and Caloric Targets by Product Type

Product Type	Sweetness Intensity (vs. 6% sucrose)	Energy Density (kcal/100 mL)	Sweetener System	Comments
Hypotonic (Rapid Rehydration)	30–40%	<10	Stevia + erythritol	Crisp and refreshing; low osmolality

Isotonic (Performance)	70–100%	20–30	Glucose + maltodextrin + sucralose	Balanced energy and taste
Hypertonic (Recovery)	100–120%	35–45	Maltodextrin + sucrose + fructose	Energy-dense; supports glycogen restoration
Functional "Zero"	40–60%	<5	Sucralose + acesulfame-K or stevia + allulose	Clean sweetness, stable under heat
Clean-Label / Natural	60–80%	15–25	Cane sugar + fruit concentrate + stevia	Natural taste with calorie moderation

11.1.5 Formulation Insight

Sweetness in hydration beverages is architecture, not decoration. A well-constructed profile supports rapid absorption in hypotonic formats, sustains energy without gastric drag in isotonic systems, and signals nourishment in recovery blends. The most effective sweetness systems are layered: a fast-onset top note (fructose, erythritol) to mask initial saltiness, a mid-body core (allulose, isomaltulose) for roundness and osmotic neutrality, and a clean, lingering tail (stevia Reb M, monk fruit) to avoid crash or cloying aftertaste. Crucially, sweetness must be calibrated after electrolytes and acids are fixed, ionic load shifts perception dramatically. When reducing sugar, compensate not just with high-intensity sweeteners, but with texture (glycerol, low-acyl gellan) and aroma lift (citrus top notes) to preserve the feeling of refreshment. In clean-label systems, this triangulation, sweetness, acid brightness, and mouthfeel, is what turns functional necessity into repeatable pleasure.

- Sweetness balance defines perceived refreshment, too sweet suppresses thirst drive; too weak makes the product feel flat.

- Always align sweetness curve with product intent: fast and clean for rehydration, round and lasting for recovery.
- Natural sweeteners (stevia, monk fruit, allulose) work best when supported by mild acids (citric, malic) that sharpen perception and mask off-notes.
- Conduct time–intensity sensory profiling to fine-tune onset and linger curves.
- When reducing sugar, compensate for body and flavor release using polyols or soluble fibers (inulin, polydextrose).

11.2 Designing Sweetness for Functional Performance and Clean Label Appeal

Sweetness in hydration and performance beverages must do more than please the palate, it should support physiological function, enhance perceived freshness, and align with regulatory and consumer expectations for transparency and natural sourcing. Modern formulations aim for "functional sweetness", where the choice of sweetener system contributes to energy delivery, absorption rate, and metabolic comfort, while maintaining a clean, natural label.

11.2.1 Functional Sweetness and Performance Integration

Hydration beverages operate within precise physiological windows. Overly sweet formulations can slow gastric emptying and impair rehydration, while under-sweetened ones may reduce palatability and energy delivery. The formulator's goal is to align sweetness intensity with activity purpose and metabolic demand.

Table 11.2.1: Functional Sweetness and Their Functional Contribution

Application	Desired Function	Sweetness Range (vs. sucrose = 100%)	Ideal Sweetener Base	Functional Contribution
Rapid Hydration (Hypotonic)	Fluid replacement, low GI	25–40%	Stevia + erythritol	Promotes fast absorption, zero-calorie

Endurance Performance (Isotonic)	Sustained energy and electrolyte balance	60–100%	Glucose + maltodextrin	Maintains plasma glucose, supports muscle glycogen
Recovery (Hypertonic)	Replenishment and glycogen restoration	100–120%	Maltodextrin + fructose	Rapid carbohydrate resynthesis
Mental & Focus Drinks	Cognitive clarity, no crash	40–60%	Allulose + stevia	No glycemic spike, long sensory tail
Clean-Label Functional	Natural appeal, moderate sweetness	50–70%	Cane sugar + fruit extract	Familiar taste, consumer trust

Performance Note:

The glucose: fructose ratio (2:1) is often ideal in isotonic systems, optimizing carbohydrate oxidation rates during endurance activity without causing GI distress.

Metabolic and Sensory Harmony

From a physiological standpoint, not all sweetness contributes equally:

- Glucose and maltodextrin support fast energy turnover, ideal during sustained effort.
- Fructose complements this by enhancing hepatic glycogen storage post-exercise.
- Allulose and erythritol provide body and mild sweetness without caloric burden or osmotic interference.
- Stevia or monk fruit contribute the sensory signature of sweetness without metabolic load.

Combining fast-metabolizing sugars with low-calorie or natural modifiers yields balanced sweetness curves that maintain both refreshment and endurance comfort. For instance:

6 g glucose + 3 g maltodextrin + 0.01% stevia provides a perceived sweetness equivalent to 6% sucrose, with superior hydration kinetics and reduced calorie density.

11.2.2 Designing for Clean Label and Regulatory Simplicity

The modern hydration consumer expects transparency, simplicity, and minimal artificial additives. Clean-label sweetness design means replacing synthetics and high-intensity chemicals with recognizable or naturally derived alternatives that maintain taste and stability.

Clean-Label Sweetening Strategies:

1. Natural high-intensity sweeteners:
   - Stevia (Reb-M, Reb-D): clean taste, low bitterness, pH-stable.
   - Monk fruit extract (mogrosides): heat-stable, rounded sweetness, no calories.
2. Rare sugars and polyols:
   - Allulose and erythritol for sugar-like texture, minimal glycemic effect.
   - Often paired with natural acids to sharpen taste perception.
3. Fruit-derived sweetness systems:
   - Apple, pear, or date extracts contribute natural sugars and body, useful for "juice-based electrolyte waters."
   - Pair with citrus acids and trace salt to prevent excessive sweetness buildup.
4. Reducing total sugar load:
   - Replace part of sucrose with maltodextrin or soluble fiber to retain mouthfeel without affecting osmolality.
   - For "reduced sugar" claims, ensure a minimum 30% sugar reduction vs. standard formula.

Clean-Label Sweetener Systems and Functional Behavior:

Table 11.2.2: Clean-Label Sweetener Systems and Functional Behavior

System Type	Composition	Calories (kcal/100 mL)	pH / Thermal Stability	Mouthfeel	Clean-Label Suitability	Notes
Stevia + Erythritol Blend	0.02% stevia + 2% erythritol	5	Excellent	Light, crisp	✓✓✓	Ideal for sparkling and clear hypotonic drinks
Allulose + Fruit Juice Base	3% allulose + 10% juice	15–20	Good	Full, natural	✓✓	Supports "no added sugar" positioning
Cane Sugar + Stevia Hybrid	3% sugar + 0.01% stevia	12	Excellent	Round, familiar	✓✓✓	Balances natural taste and calorie reduction
Glucose + Maltodextrin + Stevia	4% total carbs + 0.008% stevia	18	Excellent	Smooth, mild	✓✓	Performance-oriented isotonic blends
Fructose + Monk Fruit	2% fructose + 0.005% monk fruit	10	Very good	Clean, fruity	✓✓	Ideal for cognitive and light functional drinks

11.2.3 Optimizing Sweetness Perception with Acids and Electrolytes

Sweetness intensity is strongly modulated by acidity and ionic strength:

- Mild acidity (citric, malic) enhances brightness and refreshment perception, reducing required sweetener level by up to 20–30%.
- Sodium ions at low levels (50–100 mg/L) can round out bitterness and sharpen sweetness.
- Potassium ions, while beneficial for hydration, may introduce slight bitterness, mask with citrus or berry flavor profiles.

Practical Tip:

A formulation with 0.2% citric acid and 0.05% sodium citrate will taste sweeter and fresher at the same sugar level than one with sugar alone, due to pH-driven enhancement of sweetness and volatile release.

11.2.4 Processing and Stability Considerations

- Thermal Stability: Ensure sweeteners (especially stevia or monk fruit) are added after pasteurization to prevent degradation.
- Mixing Sequence: Dissolve bulk carbohydrates first; add high-intensity sweeteners and acids last, under gentle agitation.
- Shelf Stability: High-intensity natural sweeteners are resistant to microbial growth but sensitive to oxidation, pair with antioxidants (ascorbic acid 0.01–0.02%).
- Homogeneity Testing: Use refractive index (°Brix) checks at multiple points in the production tank to confirm even distribution.

11.2.5 Formulation Insight

- Build sweetness as a structural layer, not a single ingredient. Use complementary sweeteners to control onset, linger, and body.

- Always align sweetness density with hydration speed, lighter for rapid rehydration, richer for recovery systems.
- Emphasize natural source declaration (e.g., "sweetened with stevia and fruit sugars") to strengthen clean-label appeal.
- Test sensory balance under chilled (4°C) and ambient (25°C) conditions, sweetness perception decreases by ~15% when served cold.
- Document all sweeteners' regulatory status (e.g., GRAS, EFSA-approved) to streamline compliance for global distribution.

12. Flavor Design and Aroma Release Systems

Flavor is the emotional center of a hydration beverage. While electrolytes, carbohydrates, and stabilizers determine physiological function, it is flavor that defines consumer loyalty and brand identity. In hydration systems, where saltiness, acidity, and sweetness must coexist, flavor design becomes an act of balance and purpose.

This chapter explores how to construct flavor systems that reinforce product function, enhance sensory refreshment, and remain stable throughout processing and storage. It emphasizes the science of flavor layering, functional pairing, and controlled aroma release to create hydration beverages that feel as good as they perform.

12.1 Flavor Layering and Functional Pairing Principles

Flavor layering is the technique of combining multiple flavor notes, top, middle, and base, to achieve complexity, depth, and functional harmony in a beverage. In hydration formulations, it serves both technical (masking, balancing, solubility) and psychological (expectation, refreshment, memory) purposes. A well-layered flavor system transforms electrolytes and acidity from sensory challenges into appealing taste experiences.

12.1.1 Understanding Flavor Architecture

A functional flavor profile can be visualized as a three-tier structure:

Table 12.1.1: Functional Flavor Profile Three-tier Structure

Flavor Layer	Sensory Role	Typical Compounds or Notes	Duration	Design Objective
Top Notes	First impression; perceived immediately	Citrus esters (lemon, lime, grapefruit), mint, cucumber	Short (0–3 s)	Deliver freshness, announce the flavor identity

Middle Notes	Body and character	Berry, tropical fruit, peach, tea, floral elements	Medium (3–10 s)	Build recognition and naturalness
Base Notes	Aftertaste, depth, and longevity	Vanilla, spice, ginger, subtle herbal tones	Long (10–30 s)	Provide finish, smoothness, and memorability

The combination and balance of these layers determine how the beverage is perceived from the first sip to the lingering aftertaste. A product that fades too fast feels "watery," while one that lingers excessively can conflict with electrolytic or acidic tones.

12.1.2 Flavor Families and Functional Pairing Strategies

Each beverage type benefits from a specific flavor family designed to complement its electrolyte and sweetness profile:

Table 12.1.2: Flavor Families and Functional Pairing

Product Type	Core Flavor Families	Functional Behavior	Rationale
Hypotonic (Rapid Rehydration)	Lemon–lime, cucumber–mint, green apple, coconut water	Clean, volatile, fast-release	Supports thirst-quenching perception and crisp finish
Isotonic (Performance)	Citrus–berry, tropical punch, passionfruit–mango	Balanced sweetness and acidity	Complements carbohydrate mouthfeel; masks saltiness
Hypertonic (Recovery)	Banana, vanilla, cocoa, caramel	Soft, warm, round	Enhances comfort and post-exercise satisfaction

Functional / Mental Focus	Grapefruit, lychee, white tea, ginseng	Aromatic, slightly bitter	Reinforces cognitive alertness; balances mild stimulant tones
Electrolyte Water / Minimalist	Lemon peel, cucumber, mineral, floral	Delicate, near-transparent	Suggests purity and hydration without overt sweetness

Design Note: A successful hydration flavor must "invite drinking", high refreshment value encourages voluntary fluid intake, which improves actual hydration efficacy in performance contexts.

12.1.3 Functional Pairing with Acids and Electrolytes

Electrolytes can alter the perception of flavor molecules through ion–flavor interactions.
Understanding these interactions allows for smarter flavor design:

- Sodium (Na^+): Enhances perceived sweetness and balances sourness; brightens citrus flavors.
- Potassium (K^+): Slightly bitter at high levels; pairs best with berry, tropical, or herbal flavors that naturally mask bitterness.
- Magnesium (Mg^{2+}): Metallic taste; offset with cocoa, vanilla, or savory umami undertones.
- Calcium (Ca^{2+}): Chalky or dry; counter with creamy or acidic fruit bases (peach, apricot, orange).

Acid systems also shape flavor tone:

- Citric acid: sharp, bright, refreshing, ideal for isotonic and hypotonic systems.
- Malic acid: smoother sourness with longer persistence, good for apple, berry, and tropical notes.
- Lactic acid: mild and rounded, complements recovery or dairy-style beverages.

- Tartaric acid: sharp and crystalline, enhances clarity in citrus and sparkling beverages.

By pairing electrolyte type with acid system and flavor family, formulators can control both sensory profile and physiological perception.

12.1.4 Building Layered Flavor Systems

Step 1: Define Functional Target

Identify whether the beverage's goal is rapid hydration, sustained performance, or recovery.
Each requires different sensory pacing: light and quick for hydration; dense and round for recovery.

Step 2: Select Core Flavor Backbone

Choose 1–2 middle notes that define the flavor identity (e.g., mango–passionfruit, lemon–lime).

Step 3: Add Top Notes for Freshness

Incorporate volatile esters or oils (citral, limonene, linalool) for immediate sensory impact.

Step 4: Anchor with Base Notes

Add a subtle, low-volatility component (vanilla, ginger, tea) to improve longevity and mouth-coating balance.

Step 5: Adjust with Modifiers

Fine-tune acidity, sweetness, and salinity until the flavor supports both technical function and emotional appeal.

Table 12.1.4: Example Flavor Layering Models for Functional Beverages

Beverage Type	Top Notes	Middle Notes	Base Notes	Supporting Elements	Sensory Goal

Rapid Hydration	Lemon, lime	Green apple	Mint	Citric acid, NaCl 80 mg/L	Crisp, clean, thirst-quenching
Endurance Performance	Orange zest	Mango, passionfruit	Pineapple	Malic acid, Na/K blend	Bright, energetic, balanced
Recovery	Vanilla	Banana	Cocoa	Lactic acid, Mg/Ca blend	Comforting, creamy, replenishing
Mental Focus	Grapefruit	White tea	Ginseng, basil	Citric + tartaric acids	Aromatic, sharp, clear
Clean-Label Electrolyte Water	Cucumber	Aloe, pear	Floral	Citric acid, low-salt	Transparent, natural purity

Flavor Modulation for Target Sensory Experience:

- Refreshing: emphasize citrus and green notes, moderate acidity, short linger.
- Comforting: emphasize creamy or vanilla base, round acidity, low astringency.
- Energizing: use bright tropical or tart berry flavors with a vivid aromatic top.
- Natural / Premium: use botanical or tea infusions with soft fruit accents; minimal artificial esters.

12.1.5 Formulation Insight

- Layering flavors allows sensory continuity, each sip feels fresh, not repetitive.
- Volatile aroma compounds should be added post-pasteurization or via encapsulation to preserve brightness.
- Electrolytes interact with flavor molecules, always test flavor stability and clarity under full ionic load.

- Maintain flavor identity consistency across formats (powder, RTD, concentrate) by adjusting acidity and top-note loading.
- Conduct triangle sensory tests during development to ensure flavor differentiation remains perceptible even after storage or carbonation.

12.2 Controlled Aroma Release and Flavor Encapsulation Systems

Aroma defines the emotional "signature" of a beverage, the first sensory cue that shapes expectation and the lingering note that drives repeat consumption. However, aroma compounds are also the most volatile and fragile elements in a formulation. In electrolyte and nutrition drinks, especially those that undergo heat treatment, carbonation, or extended shelf storage, maintaining consistent aroma intensity requires both physical encapsulation and controlled release engineering.

This section outlines how to design, protect, and deliver aroma systems that retain freshness, stability, and impact from production to consumption.

12.2.1 The Challenge of Aroma Volatility

Volatile aromatic compounds (esters, aldehydes, terpenes) are chemically reactive and easily lost through:

- Evaporation during mixing, filling, or carbonation.
- Thermal degradation during pasteurization or hot-fill.
- Oxidation via dissolved oxygen or metal ions.
- Adsorption onto bottle walls or closures (especially in PET).
- Interaction with electrolytes or acids that alter solubility or perceived intensity.

As a result, beverages often lose up to 40–60% of top-note aroma intensity within three months if not stabilized or protected.

12.2.2 Aroma Retention Mechanisms

To control aroma loss, formulators employ strategies that combine physical encapsulation, emulsion stabilization, and phase management.

Primary Techniques:

1. Emulsion Stabilization
   - Disperse volatile oils in water via oil-in-water emulsions stabilized by gum arabic, modified starch, or sucrose esters.
   - Typical oil droplet size: 0.1–1.0 μm for long-term clarity and controlled release.
   - Use emulsifier blends (e.g., polysorbate 80 + gum arabic) for balanced interfacial protection.
2. Spray-Dried Encapsulation
   - Convert liquid aroma oils into stable powders using carriers such as maltodextrin, inulin, or modified starch.
   - Protects aroma during heat processing and allows easy incorporation into dry-blend or powder drink formats.
   - Rehydration releases aroma upon mixing, offering "freshly opened" sensory effect.
3. Microencapsulation / Coacervation
   - Encapsulate volatile compounds within microcapsules (10–100 μm) using gelatin–gum arabic or alginate systems.
   - Enables delayed or sustained release (e.g., aroma release upon agitation or temperature change).
   - Especially useful for recovery drinks or functional beverages with longer consumption times.
4. Inclusion Complexes (Cyclodextrins)
   - Cyclodextrins (α-, β-, or γ-type) form non-covalent "host–guest" structures that trap hydrophobic aroma molecules.
   - Release occurs gradually as the beverage is diluted, consumed, or warmed.
   - Effective for stabilizing citrus, mint, and fruity esters in electrolyte waters.

Table 12.2.2: Aroma Stabilization Technologies and Applications

Technology	Encapsulating Agent	Application Type	Thermal Resistance	Release Behavior	Clarity Impact	Notes
Oil-in-Water Emulsion	Gum arabic, sucrose ester	Isotonic, RTD drinks	Moderate	Gradual, during agitation	Slight haze possible	Best for clear isotonic and flavored waters
Spray-Dried Powder	Maltodextrin, modified starch	Dry blends, instant powders	High	Fast, upon rehydration	Clear after mixing	Ideal for sachets or powders
Coacervated Microcapsules	Gelatin–gum arabic, alginate	Protein or dairy-based recovery drinks	High	Delayed, body heat or shaking	Opaque	Suitable for enriched nutrition drinks
Cyclodextrin Complexes	β-cyclodextrin	Clean-label or premium hydration	Excellent	Slow, temperature-activated	Clear	Protects sensitive top notes

Liposomal Encapsulation	Phospholipids, lecithin	Functional or nutraceutical beverages	Moderate	Sustained, bioavailable release	Slight opalescence	Useful for aroma + bioactive co-delivery

12.2.3 Engineering Controlled Aroma Release

Controlled aroma release ensures the beverage delivers consistent freshness at every sip, even after prolonged storage. This is achieved through matrix tuning and trigger-based release:

- Physical triggers: agitation, carbonation pressure drop, temperature rise (as beverage warms during drinking).
- Chemical triggers: pH shift, ionic exchange, or dilution that alters encapsulation equilibrium.
- Temporal release design: adjusting capsule wall thickness or emulsifier layer density to control release rate.

Example:

A lemon–lime flavor encapsulated in β-cyclodextrin (0.15%) releases volatile esters more gradually, maintaining perceived freshness up to 9 months, compared to 3 months in non-encapsulated systems.

12.2.4 Flavor–Electrolyte Compatibility Considerations

Electrolytes influence both aroma solubility and volatility:

- Sodium ions enhance ester volatility, boosting aroma intensity.
- Potassium and calcium reduce volatility by increasing water structuring, can dull aroma perception.

- Citric acid brightens citrus aromas by forming soluble complexes with terpene alcohols.
- High ionic strength (>2%) can "salt out" volatile compounds; mitigate by using emulsions or co-solvents (e.g., propylene glycol <0.2%).

Practical Adjustment:

If flavor fade is detected over time, reformulate with 10–20% more top-note load or apply encapsulation through gum arabic-based emulsions for stability without altering taste.

12.2.5 Stability Testing for Aroma Systems

Aroma systems are among the most delicate components in a hydration beverage, highly sensitive to oxidation, volatilization, and interactions with acids, electrolytes, and emulsifiers. Stability testing provides the framework for evaluating how these aromatic compounds behave under heat, light, storage, and processing stress, revealing vulnerabilities that can compromise flavor impact over time. By systematically assessing these factors, formulators gain the insight needed to build aroma systems that remain vivid, balanced, and reliable throughout the product's shelf life.

Table 12.2.5: Stability Testing Protocols for Aroma Systems in Functional Hydration Beverages

Test Type	Objective	Recommended Conditions	Evaluation Metric
Headspace GC Analysis	Measure aroma retention	25°C and 40°C storage	% volatile recovery over time
Sensory Retention Panel	Evaluate perceived freshness	0, 3, 6, 9 months	Hedonic scale (1–10) for aroma intensity
Thermal Stress Test	Simulate pasteurization impact	85°C for 30 sec	Aroma loss <15% acceptable

Packaging Interaction	Assess aroma adsorption	PET, glass, aluminum	Mass transfer coefficient (K) monitoring
Carbonation Release Test	Observe volatile release	3–4 g/L CO_2	Aroma burst timing and decay rate

12.2.6 Formulation and Processing Guidelines

- Always add aroma oils post-homogenization or at the final mixing stage under low shear to minimize volatilization.
- For hot-fill systems, incorporate encapsulated or emulsified aroma that withstands up to 90°C without phase separation.
- In carbonated beverages, avoid high surface activity agents (proteins, gums) that can strip volatile compounds.
- Deaerate mixing water to reduce oxidative degradation of esters and aldehydes.
- Verify emulsion particle size and zeta potential, stable systems maintain >30 mV charge and <1 µm droplet size.

12.2.7 Formulation Insight

- Encapsulation extends aroma half-life up to 3× longer, ensuring consistent consumer experience throughout shelf life.
- Aroma design must consider release rate during drinking, too rapid feels artificial, too slow feels muted.
- Always validate sensory performance after heat processing and carbonation; encapsulation efficacy often declines under real-world stress.
- For premium hydration products, cyclodextrin complexes offer unmatched stability and clarity with minimal ingredient declaration complexity.
- Conduct accelerated light and oxygen exposure tests to forecast long-term aromatic performance before scaling.

12.3 Flavor Stability, Oxidation Control, and Aroma Preservation Technologies

Flavor stability determines how well a hydration beverage maintains its sensory identity over time. Even the most beautifully layered aroma system can deteriorate if oxidation, light exposure, or ionic reactions break down its delicate volatile compounds. Flavor degradation not only reduces freshness but can also generate "off-notes", cardboard, metallic, or cooked flavors, that quickly erode consumer trust.

Achieving long-term flavor integrity requires an integrated approach that combines ingredient protection, formulation chemistry, and packaging design to safeguard aroma from the moment of production through the final sip.

12.3.1 Mechanisms of Flavor and Aroma Degradation

Flavor deterioration occurs through several intertwined mechanisms:

1. Oxidation of Volatile Compounds
    - Citrus and fruit esters oxidize to aldehydes and acids, creating dull or sour notes.
    - Terpenes (e.g., limonene) form peroxides that impart bitterness or waxy off-flavors.
    - Oxygen ingress through packaging is a major driver.
2. Light-Induced Degradation (Photo-Oxidation)
    - UV and visible light trigger breakdown of colorants and aroma molecules, especially carotenoid-based pigments and essential oils.
    - Even clear beverages stored under retail lighting can lose up to 30% of aroma intensity in two months.
3. Electrolyte and Acid Reactions
    - High salt concentrations accelerate ester hydrolysis, reducing aroma intensity.

- Strong acids can alter volatile balance, especially in citrus and tropical profiles.

4. Metal-Catalyzed Reactions

   - Trace iron, copper, or manganese catalyze oxidative chain reactions, even at ppm levels.
   - Common sources: process water, stainless steel contact, or colorant residues.

5. Thermal Stress During Processing

   - Pasteurization or hot-fill (80–90°C) volatilizes and oxidizes aroma compounds if unprotected.
   - Emulsion droplet coalescence reduces effective surface area for aroma release.

12.3.2 Core Strategies for Flavor and Aroma Preservation

To preserve aroma integrity, flavor stability must be engineered at multiple levels, molecular, physical, and packaging.

1. Antioxidant Systems:

Antioxidants prevent or delay oxidative reactions by scavenging free radicals or chelating transition metals.
Common systems include:

Table 12.3.2A: Key Antioxidants and Their Functions

Antioxidant	Functional Role	Typical Level	Solubility	Notes
Ascorbic Acid (Vitamin C)	Primary oxygen scavenger	100–200 ppm	Water-soluble	Effective but can degrade colorants at high pH
Tocopherols (Vitamin E)	Lipid-phase antioxidant	50–150 ppm	Oil-soluble	Best for emulsified flavors or microcapsules

Rosemary Extract	Natural phenolic antioxidant	50–100 ppm	Oil-soluble	Clean-label, improves citrus stability
Green Tea Polyphenols	Antioxidant and flavor synergist	100–200 ppm	Water-soluble	Adds mild herbal tone; complements natural positioning
EDTA / Sodium Citrate	Metal chelators	10–50 ppm	Water-soluble	Prevents metal-catalyzed oxidation

Best Practice: Combine ascorbic acid + rosemary extract + citrate buffer for balanced aqueous and lipid-phase protection.

2. Oxygen Management and Deaeration:

Dissolved oxygen is the most critical variable influencing flavor stability:

- Maintain <1.5 mg/L dissolved O_2 after blending.
- Apply vacuum or nitrogen sparging during mixing to strip oxygen from water and syrup.
- Use nitrogen headspace flushing immediately before capping to displace residual oxygen.
- For powders or concentrates, store under inert gas (N_2 or CO_2) to prevent oxidation during storage.

3. pH and Ionic Control:

pH affects both oxidation rate and ester stability:

- Maintain pH 3.2–3.8 for most fruit and citrus systems, optimal for both flavor retention and microbial stability.
- At higher ionic strength, use citrate or lactate buffers to minimize hydrolysis.
- Avoid chloride-heavy salt systems (NaCl >1%) in delicate flavor bases, replace with sodium citrate or potassium lactate to reduce reactivity.

4. Encapsulation and Flavor Carriers:

Flavor carriers such as propylene glycol, glycerol, or triacetin improve volatility control and solubility of aromatic oils.
Encapsulated systems (see Section 12.2) further enhance oxidative resistance by isolating volatiles from reactive environments.
Spray-dried flavor bases typically retain 80–90% aroma content after 6 months, compared to 40–50% for non-encapsulated oils.

5. Packaging Synergy:

Packaging is the final, and often decisive, layer of flavor protection.

Table 12.3.2B: Suitability of various Packaging Materials for Flavor Protection

Packaging Material	Oxygen Transmission Rate (cc/m²/day)	UV Resistance	Suitable For	Notes
Amber Glass	0	Excellent	Premium or sensitive beverages	Complete barrier; heavy and costly
Aluminum Can	0	Excellent	Carbonated electrolyte drinks	Best oxygen and light barrier
Multilayer PET (with UV additive)	0.02–0.05	Moderate	Isotonic, natural drinks	Lightweight; add UV absorber for natural flavors
Tetra Pak / Aseptic Carton	<0.01	Excellent	Clean-label, juice-infused hydration	Optimal for long shelf life and light-sensitive pigments

Tip: For clear products requiring visibility, use UV-absorbing PET (380–420 nm cutoff) combined with oxygen scavenger liners or foil seals to maintain both transparency and protection.

Table 12.3.2C: Flavor Stability Optimization Matrix

Instability Factor	Primary Cause	Control Strategy	Verification Test
Oxidation	Oxygen ingress	Deaeration, antioxidants	Peroxide value, GC analysis
Photo-degradation	UV exposure	UV-blocking packaging	Accelerated light test (1000 lux, 3 days)
Ester Hydrolysis	Low pH + heat	Buffer control, encapsulation	GC-MS ester retention
Metal-catalyzed reactions	Fe, Cu ions	Chelators (EDTA, citrate)	ICP metal analysis
Flavor loss in PET	Aroma adsorption	Barrier coating, cyclodextrins	Headspace GC retention
Aroma fade in carbonated drinks	CO_2 stripping	Pressure-balanced filling	CO_2 release and volatile profiling

12.3.3 Integrated Shelf-Life Design Approach

To achieve consistent flavor retention across commercial conditions:

1. Map the oxidation pathways of the chosen flavor oils and acids.
2. Select antioxidants and buffers based on pH and ionic load.
3. Validate through accelerated storage (40°C, 10 days ≈ 3 months at ambient).
4. Cross-test packaging formats under light and temperature cycling.
5. Perform sensory benchmarking every 30 days during real-time storage to monitor “freshness decay.”

Performance Target:

Maintain ≥85% aroma intensity and <5% off-note development over the product's intended shelf life (typically 9–12 months RTD; 24 months powder).

12.3.4 Formulation Insight

- Flavor degradation is cumulative, addressing oxygen, light, and metals simultaneously multiplies shelf-life performance.
- Natural antioxidant systems can replace synthetics with minimal sensory tradeoff if dosed correctly.
- Avoid over-acidification: excessive acid accelerates ester cleavage and dulls top notes.
- Deaeration and packaging often yield greater aroma stability gains than additional antioxidant load.
- Validate real-world performance under distribution conditions, not just lab tests, to ensure flavor durability across climate zones.

13. Acidity Management and pH Control Systems

Acidity is one of the most powerful levers in beverage formulation. It shapes taste, preservation, electrolyte stability, and product safety, influencing both how a drink feels on the palate and how it performs physiologically. For hydration and nutrition beverages, where ionic balance and absorption rate are critical, acidity must be precisely tuned, not too high to irritate the stomach, but sufficient to support microbial stability and flavor brightness.

This chapter explores the dual technical and sensory roles of acids and buffering systems, guiding formulators in choosing the right acid types, concentrations, and combinations for specific beverage functions and shelf-life goals.

13.1 Functional Role of Acids in Hydration Beverages

Acids perform far more than flavor adjustment; they act as multifunctional agents in beverage systems, modifying pH, stabilizing electrolytes, enhancing sweetness, preserving color, and improving product safety.

In hydration formulations, acidity contributes to three critical dimensions:

1. Sensory Balance, creating a refreshing, bright taste that complements sweetness and salt.
2. Physiological Compatibility, optimizing gastrointestinal comfort and mineral absorption.
3. Chemical and Microbial Stability, maintaining system integrity and preventing spoilage.

13.1.1 Sensory and Flavor Enhancement

Acidity provides the "spark" of refreshment that defines most electrolyte and isotonic beverages. Proper acid balance can sharpen flavors, reduce perceived sweetness, and mask metallic or bitter notes from salts or minerals.

- Citric acid produces a sharp, bright sourness that enhances citrus and fruit profiles.
- Malic acid gives smoother; lingering tartness suited to berry or apple-based systems.
- Lactic acid adds roundness and mild tang, ideal for recovery or dairy-style drinks.
- Tartaric acid imparts a crystalline, dry sourness that enhances sparkling or premium formulations.

The goal is to achieve a harmonized acid–sweet–salt triangle: the three flavor axes that together create balanced palatability and drinking motivation.

13.1.2 Physiological and Functional Roles

Acidity directly affects hydration kinetics and mineral utilization:

- Gastric emptying rate: Moderate acidity (pH 3.5–4.2) supports optimal fluid absorption; excessive acidity (<3.0) can slow gastric clearance.
- Electrolyte absorption: Organic acids (citric, malic, gluconic) form soluble complexes with Na^+, K^+, Ca^{2+}, and Mg^{2+}, enhancing bioavailability and reducing precipitation.
- Muscle buffering: Citrate and malate act as mild metabolic buffers, delaying fatigue during prolonged exercise.
- Digestive comfort: Balanced acidity prevents nausea during high-intensity activity or hot climate conditions.

13.1.3 Preservation and Microbial Control

Acidification is a key pillar of beverage microbiological safety:

- Most spoilage organisms cannot survive below pH 4.2, allowing reduced preservative load or natural preservation claims.
- Acid systems act synergistically with mild preservatives (potassium sorbate, sodium benzoate) to prevent yeast and mold growth.

- Organic acids (citric, malic, lactic) contribute to natural self-preservation by lowering water activity (a_w) and inhibiting microbial metabolism.

Target pH Ranges by Beverage Type:

Table 13.1.3A: pH Ranges for Various Preservation Strategies

Beverage Type	Target pH	Main Acids	Preservation Strategy
Hypotonic (Rapid Hydration)	3.4–3.8	Citric, malic	Cold-fill with sorbate or low-pH aseptic process
Isotonic (Performance)	3.3–3.7	Citric, malic, tartaric	Hot-fill or mild pasteurization
Hypertonic (Recovery)	3.8–4.2	Lactic, citric	Cold-fill or refrigerated distribution
Functional / Clean-Label	3.5–4.0	Citric, gluconic	Natural acids + mild preservative or UV process
Protein or Dairy-Based	4.0–4.5	Lactic, acetic, malic	pH-controlled fermentation or aseptic filling

Table 13.1.3.B: Characteristics of Common Acids Used in Hydration Beverages

Acid	Taste Profile	Buffer Capacity	Mineral Compatibility	Thermal Stability	Clean-Label Status	Notes
Citric Acid	Sharp, bright, refreshing	Moderate	Excellent with Na^+, K^+	High	✓✓✓	Most versatile; enhances sweetness perception

Malic Acid	Smooth, fruity, lingering	High	Good with Ca^{2+}, Mg^{2+}	High	✓✓✓	Best for apple, berry, and tropical notes
Lactic Acid	Mild, rounded, slightly dairy	Moderate	Good with Ca^{2+}	Excellent	✓✓	Adds body; ideal for recovery beverages
Tartaric Acid	Sharp, crystalline, dry	Low	Moderate	High	✓✓	Best for sparkling or citrus formats
Acetic Acid	Vinegar-like, strong	High	Poor	Excellent	✓	Limited to functional or savory products
Gluconic Acid	Soft, neutral	Moderate	Excellent with Ca^{2+}, Mg^{2+}	Good	✓✓✓	Excellent stabilizer for mineral fortification
Phosphoric Acid	Clean, fast, dry	High	Excellent	High	-	Used in energy drinks (not clean-label)

13.1.4 Buffering Systems and pH Control

Buffers maintain stable pH across time, temperature, and formulation stress. They are essential for avoiding flavor drift, precipitation, and microbial risk.

Common Buffer Systems:

- Citric–Citrate System: Balances acidity in fruit-based drinks; widely used for isotonic formulations.
- Malic–Malate System: Smooths sourness, stabilizes magnesium and calcium salts.
- Lactic–Lactate System: Maintains gentle acidity in recovery or protein-enriched beverages.
- Phosphate System: Provides high buffering for energy drinks or high-protein formulas.
- Carbonate–Bicarbonate System: Limited use due to instability in acidic conditions; suitable only for functional or alkaline waters.

Formulation Ratio Example (Citric–Citrate Buffer):

For pH 3.5–3.6 → Use 0.25% citric acid + 0.10% sodium citrate
Adjust slightly (±0.02%) depending on ionic load and sweetness balance.

13.1.5 Interactions Between Acids and Electrolytes

Acid–electrolyte compatibility defines clarity and taste:

- Sodium citrate enhances salt smoothness and reduces harshness.
- Potassium citrate softens bitterness and improves hydration performance.
- Calcium lactate and magnesium gluconate offer soluble mineral sources without haze formation.
- Avoid calcium phosphate or magnesium sulfate in acidic drinks, they cause turbidity and metallic aftertaste.

Formulator's Shortcut:

Combine citric acid + sodium citrate at a 2.5:1 ratio to balance sourness, buffer pH, and minimize salt harshness.

13.1.6 Acidity and Sensory Perception

Acidity not only contributes taste but also affects flavor release, sweetness perception, and refreshment feel.

- A beverage at pH 3.5–3.7 delivers the most "sparkling" perception, even without carbonation.
- Higher acidity amplifies aroma volatility, increasing perceived flavor intensity.
- Mild acidity (~pH 4.0) creates a smoother, more hydrating mouthfeel preferred for recovery products.
- Electrolytes suppress acidity perception slightly, always test sensory balance under full ionic load.

13.1.7 Formulation Insight

- Choose acids that align with both flavor system and function, citric for brightness, malic for smoothness, lactic for comfort.
- Always pre-dissolve acids before adding salts to avoid localized precipitation or pH shock.
- Use buffered systems for products requiring extended shelf stability or thermal processing.
- Verify pH uniformity after full blending, variations >0.1 can destabilize color or flavor.
- Conduct shelf pH tracking (initial + 3 months + 6 months) to monitor long-term stability.
- For clean-label claims, rely on citric–malic blends or fermented acidulants (gluconic, lactic) instead of phosphoric acid.

13.2 Buffer Systems, Mineral Stability, and Acidity Control in Processing

Acid and buffer systems are the backbone of electrolyte balance, flavor stability, and product safety in hydration beverages. While acids provide pH and sensory brightness, buffers preserve that environment over time and through processing stress. A well-designed buffer prevents pH drift during storage, maintains mineral solubility, and stabilizes flavor and color. In other words, buffers "lock in" the beverage's functional chemistry from factory to consumer.

This section explains how to select and optimize buffering systems, manage mineral interactions, and maintain consistent acidity through blending, heat treatment, and filling.

13.2.1 Role of Buffers in Beverage Systems

Buffers resist changes in pH when small amounts of acid or base are introduced, ensuring flavor consistency, microbial safety, and ionic stability.
They are especially critical in:

- Electrolyte beverages containing calcium or magnesium salts prone to precipitation.
- Thermally processed beverages, where pH shifts occur under heat.
- Long shelf-life isotonic products, where slow CO_2 absorption or flavor oxidation alters acidity.

A stable buffer system is the difference between a crystal-clear beverage and one that hazes or loses taste within weeks.

13.2.2 Mechanism of Buffering

Buffering capacity (β) describes how well a beverage resists pH change:

$$\beta = \frac{\Delta B}{\Delta pH}$$

where ΔB = moles of acid or base added per liter.
Beverages with higher buffering capacity maintain consistent flavor and microbial protection under processing stress.

Optimal buffer concentration typically ranges from 0.05–0.25%, depending on beverage strength and ionic complexity.

13.2.3 Common Buffer Systems and Their Characteristics

Common buffer systems form the chemical foundation that stabilizes pH, protects sensitive ingredients, and maintains flavor consistency in hydration beverages. Each buffer, whether citrate, phosphate, lactate, or amino acid–based, brings distinct strengths in terms of pH range, mineral compatibility, flavor impact, and regulatory acceptance. Understanding these characteristics allows formulators to select buffering systems that not only control acidity but also support clarity, nutrient stability, and overall product performance under real-world processing and storage conditions.

Table 13.2.3: Comparison of Buffer Systems for Hydration and Nutrition Beverages

Buffer System	Acid–Base Pair	Optimal pH Range	Ionic Compatibility	Typical Use (%)	Notes
Citric–Sodium Citrate	Citric acid / $Na_3C_6H_5O_7$	3.0–6.2	Excellent with Na^+, K^+	0.15–0.30	Widely used; balances flavor and salt smoothness
Malic–Sodium Malate	Malic acid / $Na_2C_4H_4O_5$	3.2–4.8	Excellent with Mg^{2+}, Ca^{2+}	0.10–0.25	Stabilizes mineral fortification and fruity flavors

Lactic–Sodium Lactate	Lactic acid / $NaC_3H_5O_3$	3.5–4.5	Good with Ca^{2+}	0.10–0.30	Ideal for recovery or dairy-type beverages
Phosphate Buffer	NaH_2PO_4 / Na_2HPO_4	5.8–7.2	Excellent with proteins	0.10–0.20	Used in high-protein or energy drinks
Gluconic–Sodium Gluconate	Gluconic acid / $NaC_6H_{11}O_7$	3.2–4.8	Excellent with multivalent cations	0.10–0.25	Clean-label, mineral-friendly system
Acetic–Sodium Acetate	Acetic acid / $NaC_2H_3O_2$	3.5–5.0	Moderate	0.05–0.15	Used in savory or functional beverages

13.2.4 Mineral Stability and Precipitation Control

Electrolyte systems are sensitive to pH and ionic balance. Incorrect buffer selection can lead to precipitation, turbidity, or flavor drift.
Understanding mineral solubility curves allows precise control of clarity and shelf life.

Key Interactions:

- Calcium and Phosphate: Precipitate below pH 5.2, avoid phosphate buffers with calcium fortification.
- Magnesium and Sulfate: May form crystals in concentrated isotonic drinks, use lactate or gluconate forms.
- Iron and Copper: Catalyze oxidation reactions, chelate with citrate or EDTA (10–30 ppm).
- Sodium and Potassium: Usually stable, but affect taste balance, fine-tune with pH and citrate ratio.

Best Practice:

Maintain pH 3.4–3.8 for isotonic systems containing citrates and 3.8–4.2 for recovery beverages containing lactates or gluconates.

13.2.5 Process-Driven Acidity Shifts

Acid and buffer balance can drift during:

- Thermal processing: Heat increases acid dissociation and CO_2 loss, raising pH by 0.05–0.15.
- Carbonation: CO_2 dissolution lowers pH temporarily by 0.1–0.3.
- Storage: Oxidation of citric or malic acids may raise pH slightly over time.

To counter these shifts:

- Pre-adjust formulations 0.05–0.1 pH units below target before hot-fill.
- Use dual-buffer systems (e.g., citric + malic) for broader resistance.
- Verify post-process pH within ±0.05 of design value.

13.2.6 Integration with Electrolytes and Flavor Systems

Integrating buffer systems with electrolytes and flavor components requires a precise understanding of how acidity, mineral ions, and aromatic compounds influence one another. The balance between these elements determines not only pH stability but also flavor expression, clarity, and overall sensory harmony. By aligning buffer strength with electrolyte load and selecting flavor systems that remain stable within the resulting environment, formulators can create beverages that stay consistent, vibrant, and physiologically effective throughout their intended shelf life.

Acid–base balance directly influences flavor perception and electrolyte solubility. Design your buffer to support all three formulation pillars:

Table 13.2.6: Key Flavor Systems and Impact

System	Flavor Impact	Mineral Solubility	Notes

Citric–Citrate	Bright, refreshing	Excellent for Na^+, K^+	Standard isotonic system
Malic–Citrate	Fruity, smooth	Good for Mg^{2+}	For tropical and berry profiles
Lactic–Citrate	Mild, round	Moderate	Good for recovery beverages
Gluconic–Malic	Neutral	Excellent for Ca^{2+}, Mg^{2+}	Clean-label and mineral-stable

Formulation Tip:

Adding 0.1% sodium citrate can smooth perceived acidity, reduce metallic off-notes, and stabilize aroma esters in high-salt formulations.

13.2.7 pH Control in Production and Quality Assurance

Precise pH management is critical for reproducibility and regulatory compliance.

Recommended QA Controls:

1. Inline pH monitoring: Continuous sensors during blending and before filling.
2. Batch verification: Measure pH at 20°C ± 1°C using calibrated glass electrode.
3. Post-thermal check: Record pH immediately after cooling.
4. Shelf simulation test: Accelerated storage at 40°C for 14 days to confirm pH drift ≤0.1.
5. Microbial safety audit: Confirm pH <4.2 for acid-preserved systems.

Tolerances:

- pH deviation within ±0.05 of formulation target is acceptable for production scale.
- Larger deviation may indicate acid loss, incorrect buffer ratio, or incomplete mixing.

13.2.8 Buffer Optimization Workflow

1. Define target pH based on flavor and preservation needs.
2. Select acid pair (citric, malic, lactic, etc.) aligned with flavor system.
3. Add buffer base (sodium, potassium, or calcium salts) gradually under mixing.
4. Fine-tune with ionic load simulation, replicate full electrolyte composition before final adjustment.
5. Validate under process conditions (temperature, carbonation, pasteurization).
6. Document buffer ratio for batch traceability and QA consistency.

13.2.9 Formulation Insight

- Buffer systems act as the "governor" of beverage chemistry, controlling both flavor brightness and mineral clarity.
- Avoid over-buffering: excessive buffering (>0.3%) can flatten acidity and dull flavor.
- Always match buffer choice to both acid type and electrolyte matrix to ensure solubility and sensory alignment.
- When switching acid systems (e.g., from citric to malic), recalibrate sodium or potassium salts, their buffering strengths differ.
- Use temperature-compensated pH meters for precise readings during hot-fill or pilot trials.

13.3 Acidity and Electrolyte Perception - Sensory Balancing in Formulated Systems

The balance between acid, salt, and sweetness defines the overall sensory "architecture" of a hydration beverage. Consumers often describe successful formulations as refreshing, smooth, or clean-tasting, all qualities rooted in acidity management and ionic interactions. A precise acid–electrolyte balance ensures not only flavor harmony but also physiological functionality, optimizing both palatability and absorption.

This section explains how pH, acid selection, and mineral composition influence perceived saltiness, sweetness, and refreshment, and how formulators can tune these interactions to deliver a consistent sensory experience.

13.3.1 Sensory Chemistry of Acid–Salt Interaction

Acids and electrolytes interact on multiple sensory and molecular levels:

- Ionic Masking: Acidic protons (H^+) reduce the perceived harshness of sodium and potassium salts by interfering with taste receptor activation.
- Charge Competition: Hydrogen ions compete with cations (Na^+, K^+, Ca^{2+}) for receptor binding, muting salt intensity.
- Flavor Enhancement: Acidity amplifies volatile release, making flavors more vibrant and aromatic.
- Sweetness Modulation: Moderate acidity enhances perceived sweetness; excessive acidity suppresses it.

The relationship follows a curved response, there is a "sweet spot" of pH (~3.5–3.8) where acidity, sweetness, and salinity reach equilibrium, producing optimal refreshment perception.

13.3.2 pH–Taste Interaction Matrix

Table 13.3.2: pH–Taste Interaction Characteristics and Applications Matrix

pH Range	Sensory Characteristics	Recommended Application
<3.2	Sharp, sour, aggressive; sweetness muted; metallic salts intensified	Avoid for isotonic drinks; acceptable for sports shots or energy gels
3.3–3.8	Bright, refreshing, balanced; sweetness and salt perception in harmony	Ideal for isotonic and hypotonic hydration drinks

3.9–4.3	Mild, smooth, less "sparkle"; suitable for recovery or protein beverages	Use malic/lactic acids to enhance mouthfeel
>4.4	Flat or dull; low perceived refreshment; higher microbial risk	Avoid unless cold-stored or preserved by other means

Key Point: pH control is both a flavor tool and a functional safeguard, optimize it for sensory performance, not just shelf stability.

13.3.3 Influence of Acid Type on Electrolyte Taste

Different acids interact uniquely with mineral salts, altering taste and mouthfeel. Understanding these relationships allows targeted sensory tuning.

Table 13.3.3: Influence of Key Acid Types on Electrolyte Taste

Acid Type	Impact on Electrolyte Perception	Flavor & Texture Outcome	Recommended Pairing
Citric Acid	Masks sodium harshness, enhances sweetness; brightens flavor	Crisp, lively taste	Na^+, K^+ systems
Malic Acid	Smooths bitterness, rounds mouthfeel	Soft, long-lasting tartness	Mg^{2+}, Ca^{2+} systems
Lactic Acid	Reduces metallic notes; adds creamy tang	Gentle, mild acidity	Ca^{2+} recovery drinks
Gluconic Acid	Softens high-mineral load, low impact on sweetness	Neutral, mineral-balanced	Fortified or functional beverages
Phosphoric Acid	Intensifies sharpness and carbonation "bite"	Clean, dry finish	Energy drinks, cola-type bases

13.3.4 Electrolyte–Acid Synergy in Flavor Systems

Electrolytes affect not only taste but also mouthfeel and perceived temperature. Balanced acidity enhances these cues and reduces undesirable sensory fatigue.

Table 13.3.4: Impact of Key Electrolytes in Flavor Systems

Electrolyte	Typical Impact	Acid Interaction	Sensory Strategy
Sodium	Salty, slightly harsh	Citric acid masks harshness, adds brightness	Maintain Na^+ at 20–30 mEq/L; use citrate buffer
Potassium	Bitter, metallic	Malic or lactic acids soften and round flavor	Use pH 3.6–3.8 with sweet–acid balance
Calcium	Chalky, lingering	Lactic/gluconic acids improve solubility and mouthfeel	Maintain Ca^{2+} below 5 mEq/L; pair with mild acid
Magnesium	Bitter, astringent	Malic acid smooths and balances aftertaste	Use chelated forms; pair with fruity flavor base
Chloride ions	Sharp, briny	Neutralized by organic acid buffers	Combine with citric–malic system for isotonic clarity

Formulation Example:

A citrus isotonic drink (pH 3.6) with sodium citrate and malic acid offers a clean, bright taste without salt harshness, ideal for endurance hydration.

13.3.5 Designing Sensory Equilibrium: The Acid–Sweet–Salt Triangle

A hydration beverage's flavor stability can be visualized as a three-point balance between:

1. Acidity → drives refreshment and suppresses off-flavors
2. Sweetness → provides comfort and energy association
3. Saltiness (electrolytes) → reinforces body, functionality, and authenticity

Table 13.3.5: The Acid–Sweet–Salt Sensory Equilibrium

Axis	Function	Excess Effect	Correction Strategy
Acidic	Enhances refreshment	Too sour → reduces drinkability	Add buffer or blend acids (citric + malic)
Sweet	Masks acid/salt	Too sweet → cloying, low refreshment	Reduce sugar, adjust pH lower by 0.1
Salty	Adds realism, functional cue	Too strong → metallic/bitter	Mask with citric/lactic acids or flavor enhancers

For clean, functional hydration products, aim for acid–sweet–salt ratios that yield:

- Acid intensity: 4–5 (on 10-point scale)
- Sweetness: 5–6
- Saltiness: 2–3

This ratio delivers maximum perceived refreshment with minimal aftertaste.

13.3.6 Sensory Testing and Optimization

Recommended sensory design workflow:

1. Establish target pH using instrumental measurement.
2. Conduct bench tasting across ±0.1 pH increments to identify ideal sensory balance.
3. Adjust acid blend (citric vs. malic vs. lactic) to fine-tune mouthfeel.
4. Validate under final ionic load, electrolytes significantly alter perceived acidity.
5. Perform temperature and carbonation trials (10°C vs. 25°C, still vs. sparkling).

6. Use trained panels to quantify sourness, saltiness, and sweetness perception curves.

Typical Observations:

- Cold beverages (~10°C) show lower perceived sweetness and acidity, slightly increase acid concentration (~5%) for balance.
- Carbonation enhances perceived acidity by ~0.2 pH units, reduce actual acid load accordingly.
- Flavors high in esters (e.g., tropical fruit) tolerate lower pH (3.3–3.5); those with delicate floral notes prefer higher pH (3.8–4.0).

13.3.7 Advanced Masking and Sensory Modifiers

To perfect acid–electrolyte harmony, formulators often incorporate natural masking agents or modulators:

- Amino acids (glycine, alanine): soften saltiness and enhance sweetness synergy.
- Stevia glycosides: heighten sweetness perception at low acid levels.
- Natural bitterness blockers (e.g., thaumatin, monk fruit extract): reduce metallic or bitter off-notes.
- Glycerol or erythritol: round mouthfeel and reduce sour edge.

When combined with smart pH control, these tools allow low-sugar, clean-label hydration products to retain full flavor depth without artificial additives.

13.3.8 Formulation Insight

- The most "refreshing" drinks align acidity (citric/malic) with moderate ionic load (~40–60 mEq/L total electrolytes) and balanced sweetness (4–6°Brix).
- Potassium and magnesium often dull aroma intensity, offset with slightly brighter acidity (citric or tartaric).
- Always re-evaluate flavor and pH under final packaging conditions; PET and CO_2 retention can subtly alter acidity perception.

- For athlete-focused formulations, target pH 3.5–3.7, ideal for rapid hydration, optimal taste, and microbiological safety.
- For recovery beverages, aim for pH 3.9–4.1, smoother mouthfeel and gentler gastric tolerance.

14. Electrolyte Blending Strategies and Mineral Bioavailability

Electrolytes are the functional backbone of hydration beverages. They control osmotic balance, fluid absorption, nerve signaling, and muscle performance. However, in formulation practice, electrolytes also bring sensory, chemical, and physical challenges, from bitterness and saltiness to solubility and precipitation issues.

Designing an effective electrolyte system means achieving two simultaneous goals:

1. Deliver physiologically relevant mineral levels for hydration and recovery.
2. Maintain pleasant taste, visual clarity, and long-term stability across temperature, storage, and packaging conditions.

This chapter explores how to build synergistic electrolyte systems that satisfy both functional performance and consumer expectations.

14.1 The Science of Electrolyte Balance in Hydration Beverages

Electrolytes regulate the movement of water and solutes across cellular membranes. When consumed in correct ratios, they maintain osmotic equilibrium between intracellular and extracellular fluids, a process essential for thermoregulation, endurance, and cognitive performance.

Hydration beverages must therefore mimic or strategically adjust to the electrolyte profile of sweat, which varies with activity type, intensity, and environmental conditions.

14.1.1 Functional Roles of Key Electrolytes

Table 14.1.1: Functional Roles of Key Electrolytes

Electrolyte	Primary Physiological Role	Typical Loss in Sweat (mg/L)	Optimal Beverage Range (mg/L)	Functional Notes
Sodium (Na^+)	Maintains plasma volume, nerve transmission	800–1200	300–800	Critical for rehydration; drives thirst and absorption
Potassium (K^+)	Regulates muscle and cardiac function	200–400	150–400	Balances sodium, prevents cramping
Calcium (Ca^{2+})	Supports muscle contraction, enzyme activation	20–60	20–75	Aids recovery and signaling; stabilizes pH
Magnesium (Mg^{2+})	Energy metabolism, anti-cramp function	10–30	10–50	Enhances muscle relaxation and energy release
Chloride (Cl^-)	Maintains osmotic and acid–base balance	1000–1500	400–1000	Complements sodium; defines “saline” character
Phosphate (PO_4^{3-})	Energy transfer (ATP buffer)	trace	20–40	Used in recovery and performance formulations

Formulator’s Note:

Always balance Na^+:K^+ ratio between 2:1 and 3:1 for optimal absorption and taste. Exceeding this ratio makes drinks taste overly salty; lower ratios reduce thirst response and rehydration speed.

14.1.2 Principles of Electrolyte Synergy

Electrolytes do not act independently; their interactions define absorption rate and sensory harmony. The most successful beverages maintain a balanced osmotic pressure near that of human plasma (≈ 280–320 mOsm/kg).

Key Synergy Concepts:

- Sodium drives water uptake, while potassium maintains intracellular hydration, together, they ensure efficient rehydration.
- Citrate and lactate anions enhance mineral solubility and reduce bitterness.
- Magnesium and calcium improve long-term muscle recovery but must be stabilized with chelating acids (malic, gluconic).
- The presence of small carbohydrates (glucose, maltodextrin) accelerates sodium-linked water absorption via the SGLT1 transport mechanism.

14.1.3 Electrolyte Selection for Beverage Type

Table 14.1.3: Electrolytes for Various Beverage Ty[es

Beverage Type	Primary Function	Recommended Electrolyte Focus	Total Ionic Strength (mEq/L)	Notes
Hypotonic (Rapid Hydration)	Fast rehydration during exercise	Na^+, K^+	20–40	Low sugar; light taste, clear appearance
Isotonic (Performance)	Balanced energy and hydration	Na^+, K^+, Cl^-, small Mg^{2+}	40–80	Matches body fluid osmolality
Hypertonic (Recovery)	Post-workout replenishment	Na^+, K^+, Ca^{2+}, Mg^{2+}	80–120	Includes carbs and proteins;

				thicker mouthfeel
Functional/Wellness Waters	Mild daily support	K^+, Ca^{2+}, Mg^{2+}	10–30	Clean-label, often flavored lightly
Electrolyte Concentrates	High-efficiency rehydration	Na^+, K^+, Mg^{2+}	100–150	To be diluted before consumption

14.1.4 Enhancing Mineral Bioavailability

Mineral bioavailability depends on solubility, complexation, and carrier system. Certain acid–anion pairs form soluble complexes that increase absorption efficiency without causing turbidity.

Table 14.1.4: Mineral Bioavailability and Preferred Soluble Forms

Mineral	Preferred Soluble Forms	Bioavailability Rank	Comments
Sodium	NaCl, NaCitrate, NaLactate	★★★★★	Citrate softens saltiness and stabilizes flavor
Potassium	KCl, KCitrate, KMalate	★★★★☆	KCl often bitter; blend with malate for smoothness
Calcium	CaLactate, CaGluconate, CaCitrate	★★★★☆	Requires pH >3.8 for solubility; avoid phosphate salts
Magnesium	MgGluconate, MgMalate, MgCitrate	★★★★☆	Chelated forms reduce astringency

Chloride	NaCl, KCl	★★★★☆	Provides authenticity to "sweat-like" profile

Formulation Tip:

Use organic acid salts (lactate, citrate, gluconate) instead of sulfates or phosphates for improved clarity and absorption, particularly important in clear hydration and electrolyte waters.

14.1.5 Factors Affecting Electrolyte Stability

- pH: Below 3.2 may precipitate calcium and magnesium salts; optimal range 3.5–4.0.
- Temperature: High storage temperatures (>35°C) accelerate crystallization in supersaturated solutions.
- Total solids: Ionic strength above 100 mEq/L may cause cloudiness; dilute or buffer accordingly.
- Packaging: Avoid contact with reactive metals (tin, aluminum) without internal lacquer coating; citrates and chlorides can corrode.

Accelerated Stability Testing:

Simulate storage at 40°C for 14 days and cyclic conditions (20–40°C) to monitor haze, pH drift, and taste consistency.

14.1.6 Engineering the "Functional Osmolality Window"

Osmolality governs how quickly water moves across intestinal membranes:

- Hypotonic (150–270 mOsm/kg): Fastest absorption; best for heat conditions and athletes.
- Isotonic (280–320 mOsm/kg): Matches plasma; ideal for sustained endurance.
- Hypertonic (>320 mOsm/kg): Slower absorption; suited for recovery formulations with energy content.

Osmolality can be controlled through:

- Mineral concentration: Adjust total electrolyte load.
- Carbohydrate addition: Glucose and maltodextrin increase osmolality; adjust for hydration balance.
- Acid–base pairing: Organic acids (citrate, malate) can lower osmolality while maintaining flavor.

Performance Benchmark:

A well-balanced isotonic drink should maintain ≈290 mOsm/kg, pH 3.6–3.8, and total sodium 400–600 mg/L.

14.1.7 Practical Blending Sequence

To prevent precipitation or sensory imbalance, always follow a controlled blending order:

1. Start with deionized or filtered water (low hardness).
2. Add acids and buffers (citric, malic, or lactate) first; adjust pH.
3. Introduce sodium and potassium salts under agitation.
4. Add calcium and magnesium salts last, ensuring pH >3.6.
5. Mix until fully clear; adjust ionic ratio and flavor system.
6. Conduct clarity check (visual + turbidity <1 NTU) before filling.

Formulation Insights

- Balance physiology and flavor simultaneously: functionally correct does not always mean sensorially pleasant, buffer to achieve both.
- Sodium and potassium form the hydration base; calcium and magnesium provide recovery support and brand differentiation.
- Use blended salts (e.g., NaCitrate + KCitrate + CaLactate) for neutral taste and balanced ionic strength.
- Avoid phosphate and sulfate salts unless needed for specific energy or performance drinks.

- Adjust sweetness–salt–acid ratios after setting final mineral levels, electrolytes significantly alter perceived taste intensity.

14.2 Balancing Electrolytes for Different Activity Levels and Climate Conditions

No single electrolyte profile fits all hydration needs. The ideal composition of sodium, potassium, magnesium, and calcium varies dramatically depending on sweat rate, activity duration, intensity, and environmental temperature.

A well-formulated beverage must replicate real physiological losses while maintaining palatability and osmotic efficiency. Overly concentrated formulas can slow absorption, while insufficient electrolyte replacement leads to suboptimal rehydration or muscle fatigue.

This section provides a structured approach to designing adaptive electrolyte systems tailored to sport, climate, and performance context.

14.2.1 The Physiology of Electrolyte Loss

Sweat composition is not uniform, it changes with training adaptation, hydration state, and environment.

Table 14.2.1: Average sweat electrolyte concentrations (per liter)

Ion	Range (mg/L)	Typical Function
Sodium (Na^+)	800–1800	Regulates extracellular fluid and thirst response
Chloride (Cl^-)	1000–2000	Balances sodium and acid–base equilibrium
Potassium (K^+)	150–250	Maintains muscle excitability and nerve transmission
Calcium (Ca^{2+})	10–70	Supports muscle contraction
Magnesium (Mg^{2+})	5–40	Prevents cramping, supports metabolism

Sweat losses can exceed 2–3 L/hour under high-heat endurance conditions, which means athletes may lose 1.5–3.0 g of sodium per hour if not replenished.

14.2.2 Activity-Based Electrolyte Design

Electrolyte ratios should be adjusted to the hydration needs of specific activity categories.

Table 14.2.2: Recommended Electrolyte Composition by Activity Type

Activity Type	Sodium (mg/L)	Potassium (mg/L)	Calcium (mg/L)	Magnesium (mg/L)	Total Osmolality (mOsm/kg)	Notes
Light / Daily Hydration	100–200	100–150	10–20	5–10	150–220	Mild flavor; low osmolality; for routine consumption
Moderate Exercise (1 hr)	300–500	150–250	20–40	10–20	250–280	Isotonic profile for fitness or cycling
Endurance Training (1–3 hrs)	500–800	200–300	30–50	20–30	280–320	Classic isotonic; replace real sweat loss efficiently
High-Heat or Tropical Climate	600–1000	200–300	20–40	10–20	300–330	Slightly higher sodium to support

						high sweat rates
Post-Workout Recovery	400–600	300–400	50–75	25–40	300–350	Higher potassium and calcium; supports muscle repair
Extreme Endurance (Marathon / Triathlon)	800–1200	200–300	40–60	20–40	320–360	High sodium and citrate buffering to prevent hyponatremia

Formulator's Note: Sodium drives water retention; potassium drives cell recovery. In recovery formulations, invert the Na^+:K^+ ratio slightly (≈1.5:1) to accelerate intracellular rehydration.

14.2.3 Climate and Environmental Adjustments

Electrolyte systems must adapt not only to physical activity but also to ambient temperature and humidity.

Table 14.2.3: Climate Conditions and Adjustment Strategies

Climate Condition	Adjustment Strategy	Rationale
Hot & Humid (>30°C)	Increase sodium (by 20–25%), reduce acidity slightly	Compensates for heavy sodium loss and higher intake volumes

Dry & Arid (High Evaporation)	Increase potassium and magnesium	Dehydration accelerates intracellular imbalance
Cold Weather Sports	Reduce sodium, increase mild acids (malic/lactic)	Lower sweat rate but increased energy metabolism
Altitude (>1500 m)	Add calcium and phosphate	Supports muscular oxygen efficiency and bone metabolism
Indoor Air-Conditioned Environments	Maintain balanced isotonic ratios	Sweat loss moderate, but lower thirst cue requires flavor reinforcement

Pro Tip: The human thirst mechanism lags fluid loss; beverages with 0.3–0.5% sodium citrate not only enhance flavor but also promote faster voluntary rehydration through taste-driven feedback.

14.2.4 Adjusting Flavor and Acidity for Climate

Climate conditions also affect sensory preference and flavor tolerance.

- In hot conditions, consumers prefer higher acidity (pH 3.5–3.7) and lighter sweetness.
- In cold or indoor environments, softer acidity (pH 3.8–4.0) and smoother mouthfeel are preferred.
- Tropical profiles (mango, lime, citrus) pair best with high-electrolyte systems.
- Berry and apple flavors work well for recovery or low-sodium formats.

Formulation Example (Hot-Climate Isotonic Drink):

- pH: 3.6
- Sodium citrate: 0.35%
- Potassium chloride: 0.12%

- Magnesium lactate: 0.04%
- Malic acid: 0.10%
- Sweetness: 4.5°Brix (sucralose + glucose blend)
- Flavor: lemon–lime with 0.02% encapsulated aroma oil

14.2.5 Adaptive Electrolyte Ratios

Table 14.2.5: Adaptive Electrolyte Ratios for Target Conditons

Target Condition	$Na^+:K^+$ Ratio	$Ca^{2+}:Mg^{2+}$ Ratio	Notes
Rapid Hydration	3:1	2:1	Sodium-dominant; fast absorption
Balanced Isotonic	2.5:1	2:1	For daily or moderate training
Recovery Formulas	1.5:1	3:2	Emphasizes potassium and calcium for cellular repair
High-Heat Rehydration	3–3.5:1	2:1	Elevated sodium and chloride for sweat loss
Low-Sugar Functional Water	2:1	1.5:1	Mild ionic presence with smooth taste

Maintaining total ionic strength within 40–90 mEq/L ensures effective hydration without exceeding osmotic comfort limits.

14.2.6 Customizing for Athlete Type and Use Case

Table 14.2.6: Formulation Customization for Athlete Type and Use Case

Athlete Profile	Formulation Strategy	Key Ingredient Notes
Endurance Athlete (Cycling, Marathon)	Higher Na^+ and Cl^-; pH 3.6; isotonic carbs (4–6% glucose–maltodextrin)	Use sodium citrate and malate for sustained hydration

Strength & Recovery (Weight Training)	Higher K^+, Ca^{2+}, Mg^{2+}; lower Na^+	Add malic/lactic acids for smooth taste and muscle recovery
Heat-Exposed Worker / Military	High Na^+ (600–800 mg/L); low sweetness	Use buffered systems (citric–citrate) to avoid stomach upset
Casual Hydration (Office, Wellness)	Light electrolyte load; flavor-led	K^+ and Mg^{2+} only; mild fruit or herbal flavors
Youth / Teen Hydration	Reduced Na^+ (~200–300 mg/L); balanced K^+	Ensure mild acidity (pH ~3.9) and low total osmolality

14.2.7 Practical Design Workflow for Adaptive Electrolyte Blends

1. Define the environment and sweat loss rate (field data or literature).
2. Set target sodium and potassium range (mg/L) based on activity duration.
3. Adjust calcium and magnesium for recovery or pH buffering.
4. Confirm total osmolality (280–320 mOsm/kg for isotonic balance).
5. Validate pH (3.5–3.8 for isotonic; 3.8–4.1 for recovery).
6. Conduct sensory and performance testing (palatability + hydration markers).

Formulation Insights

- Electrolyte ratio, not just total content, defines rehydration efficiency.
- Sodium citrate and potassium malate provide optimal clarity and mouthfeel for most climate conditions.
- Always include temperature-based testing, flavor and acidity perception shift between 5°C and 25°C.

- For commercial flexibility, design modular electrolyte concentrates that can be adjusted by dilution factor for regional climates.
- When developing global SKUs, offer two regional formulations:
    - Tropical/Hot-Climate Version: higher Na^+, sharper acidity.
    - Temperate/Cool Version: lower Na^+, smoother acidity and slightly sweeter profile.

14.3 Electrolyte Carriers, Solubility Enhancers, and Clear Solution Technology

Clarity and stability are the visual proof of good formulation science. Even a perfectly balanced electrolyte drink can fail commercially if it develops haze, sediment, or clouding during storage.
Electrolyte clarity depends on ionic compatibility, solvent system, temperature tolerance, and carrier design.

This section explains how to ensure minerals remain fully dissolved and stable throughout processing and shelf life, even in high-load functional beverages.

14.3.1 Understanding Solubility Dynamics of Electrolytes

The solubility of salts in water depends on:

- pH: Certain salts (e.g., Ca^{2+}, Mg^{2+}) have narrow solubility windows.
- Ionic Strength: High total ion load can "salt out" other ions or flavor compounds.
- Temperature: Most salts dissolve better in warm water, but some (e.g., Na_2CO_3) show inverse solubility.
- Complexation: Organic acids can form soluble complexes with divalent minerals, enhancing stability.

Key principle: Solubility ≠ Stability, a salt may dissolve initially but later precipitate if ionic equilibrium shifts due to pH drift or temperature cycling.

14.3.2 Common Causes of Haze or Precipitation

Table 14.3.2: Root Causes of Haze and Precipitation in Electrolyte Beverages - Mechanisms and Prevention Strategies

Problem	Likely Cause	Corrective Strategy
White haze after blending	Insoluble calcium or magnesium salts	Switch to citrate/lactate/gluconate forms; maintain pH >3.6
Sediment after storage	Supersaturation or acid shift	Reduce ionic load; increase citrate buffer
Clouding after heat treatment	pH change or flavor–electrolyte interaction	Adjust pH pre-fill; reduce flavor oil load
Crystals in concentrate	Over-concentration or cooling	Dilute; reformulate with more soluble salt forms
Loss of transparency in PET bottle	Ionic precipitation + light scattering	Use chelators; reduce mineral total by 10–15%

Formulator's Tip: Always monitor ionic compatibility, avoid combining divalent ions (Ca^{2+}, Mg^{2+}) with phosphate, sulfate, or carbonate under acidic conditions.

14.3.3 Solubility Enhancement Strategies

To maintain full transparency and ionic stability, modern formulations rely on carrier systems and solubility enhancers.

a. Organic Acid Complexation

- Chelating acids (citric, gluconic, malic, lactic) bind with metal ions to form soluble complexes.
- These complexes remain clear across typical beverage pH (3.3–4.2).
- Example: Calcium citrate and magnesium gluconate remain fully soluble at pH ≥3.7.

b. Co-Solvent Systems

- Small polyols like glycerol (0.2–0.5%) or propylene glycol (0.1–0.3%) improve mineral dispersion and help dissolve hydrophobic flavor oils.
- Particularly useful in carbonated or high-electrolyte beverages.
- Ensure compliance with food regulations for usage levels.

c. Controlled pH Buffering

- Maintain pH stability using citric–citrate or malic–malate systems.
- Fluctuations of ±0.1 pH can cause visible precipitation in high-mineral drinks.
- Avoid unbuffered acids (e.g., direct citric acid addition) without matching base salts.

d. Temperature-Stage Solubilization

- Dissolve salts in warm water (40–50°C) during blending, then cool gradually.
- Prevents localized saturation or pH microzones that lead to crystallization.

e. Use of Chelating Agents

- Chelators (citric acid, EDTA, gluconate) bind metal ions, preventing color change and haze.
- Typical dosage: EDTA 10–30 ppm, citric acid 0.25–0.5%.
- For clean-label formulas, replace EDTA with sodium citrate + rosemary extract for oxidative stability.

14.3.4 Electrolyte Carrier Systems

Carriers provide the "vehicle" that allows minerals to dissolve evenly and remain stable through packaging and storage.

Table 14.3.4: Common Electrolyte Carriers and Their Functional Advantages

Carrier	Chemical Nature	Function in Electrolyte System	Recommended Use Level	Notes
Sodium Citrate	Organic acid salt	Buffers pH, improves sodium palatability	0.2–0.5%	Key carrier for isotonic systems
Potassium Citrate	Organic acid salt	Stabilizes potassium, reduces bitterness	0.1–0.3%	Smooth taste and clear appearance
Calcium Lactate	Calcium salt of lactic acid	Improves calcium bioavailability, minimal turbidity	0.05–0.15%	Stable at pH >3.8
Magnesium Gluconate	Organic chelate	High solubility and smooth mouthfeel	0.05–0.10%	Preferred for clear functional waters
Sodium Gluconate	Chelator and buffer	Enhances solubility of divalent ions	0.05–0.20%	Excellent for clean-label systems
Potassium Malate	Organic salt	Enhances fruit flavor and buffers pH	0.05–0.15%	Useful in tropical profiles
Calcium Citrate Malate (CCM)	Dual chelate	Sustained release and high bioavailability	0.05–0.10%	For recovery formulations

14.3.5 Clarity Optimization in Functional and Clean-Label Beverages

Achieving crystal clarity is often a key brand attribute, especially in electrolyte waters, sparkling isotonic drinks, and zero-sugar formulations.

Steps to Ensure Long-Term Clarity:

1. Start with low-mineral water (hardness <50 ppm $CaCO_3$).
2. Add acids and buffers first to set stable pH.
3. Dissolve electrolytes fully under mild heat and continuous agitation.
4. Cool before adding flavors and sweeteners.
5. Fine-filter (0.2–0.5 μm) before filling to remove undissolved particulates.
6. Use chelating buffers to maintain ionic dispersion and prevent color–metal reactions.
7. Conduct 4-week clarity stability tests under variable storage (ambient + accelerated).

Evaluation Metric:

- Turbidity <1.0 NTU (Nephelometric Turbidity Units)
- pH variation <±0.05 after 30 days
- No visible haze or sediment under 1000-lux light exposure

14.3.6 Advanced Techniques for Enhanced Clarity

a. Nanofiltration and Ion Balancing

Use nanofiltration to partially remove hardness ions from base water, reduces scaling and haze formation.

b. Microencapsulation of Electrolytes

In complex formulations (e.g., multi-functional or protein-enriched drinks), electrolytes can be microencapsulated within carbohydrate matrices or liposomes.

This prevents direct interactions that cause precipitation and allows controlled release during consumption.

c. Clarifying Agents

Natural clarifiers (e.g., gum arabic, pectin <0.05%) can be used at micro-levels to stabilize colloidal particles without visible haze formation.

d. Process Integration

Inline mixing and high-shear homogenization (100–150 bar) ensures fine ionic dispersion and prevents localized salt clustering.

14.3.7 Packaging and Storage Considerations for Clarity Retention

- PET bottles: Use UV-protective grades to prevent photochemical instability.
- Aluminum cans: Provide excellent protection against light and oxygen but require internal epoxy or BPA-free coating to prevent corrosion.
- Glass bottles: Inert and transparent; ideal for premium isotonic or wellness waters.
- Temperature control: Store finished products below 30°C to prevent micro-precipitation or flavor dulling.

Accelerated Testing Protocol:

Store samples at 5°C, 25°C, and 40°C for 14 days; evaluate pH, clarity, and taste. Any haze above 1 NTU or visible sediment denotes incompatibility.

Formulation Insights

- Electrolyte solubility is maximized through organic acid pairing and stepwise blending order.
- Avoid mixing all mineral salts simultaneously, stage by solubility and pH compatibility.
- Gluconates and lactates are excellent natural solubility enhancers for clean-label systems.

- Clarity is not just cosmetic; it's a proxy for molecular stability and perceived product quality.
- A clear drink that stays stable for 9–12 months signals robust ionic control and superior formulation engineering.

15. Carbohydrate Systems, Energy Delivery, and Mouthfeel Engineering

Carbohydrates in hydration and nutrition beverages serve multiple, interdependent roles: energy source, osmotic regulator, flavor modulator, and mouthfeel enhancer. The challenge in modern formulation is balancing these functions to provide physiological benefit without compromising clarity, palatability, or label simplicity.

From high-performance isotonic drinks to low-calorie electrolyte waters, carbohydrate design defines both the metabolic impact and consumer experience of the product. This chapter explores carbohydrate chemistry, selection, and sensory management for beverages engineered around performance, recovery, or wellness goals.

15.1 The Functional Role of Carbohydrates in Hydration and Recovery Beverages

Carbohydrates are the body's primary energy substrate during moderate to high-intensity exercise. When formulated intelligently, they accelerate fluid absorption, sustain energy availability, and influence how a beverage feels and tastes. However, inappropriate carbohydrate concentration or type can slow gastric emptying, raise osmolality beyond comfort limits, and diminish hydration efficiency.

A balanced carbohydrate system must therefore optimize energy delivery rate, hydration kinetics, and sensory satisfaction simultaneously.

15.1.1 Physiological Functions of Carbohydrates in Hydration

Table 15.1.1: Physiological Roles of Carbohydrates in Hydration and Recovery Beverages

Function	Physiological Mechanism	Formulation Implication

Energy Supply	Provides glucose for muscular ATP production	Determines carbohydrate concentration and type
Enhanced Water Absorption	Glucose–sodium co-transport via SGLT1 in intestinal lining	Requires Na^+ presence for efficient fluid uptake
Glycogen Replenishment	Post-exercise glucose restoration in muscle tissue	Relevant for recovery and hypertonic beverages
Taste and Thirst Response	Sweetness reinforces drinking motivation	Sugar type and intensity affect consumption rate
Mouthfeel and Body	Viscosity and density contribute to perceived quality	Controlled via carbohydrate chain length and ratio

15.1.2 Types of Carbohydrates Used in Beverage Formulation

Different carbohydrate structures yield distinct performance, taste, and osmotic effects. Choosing the right combination allows control over absorption rate, sweetness, and caloric density.

Table 15.1.2: Comparison of Carbohydrate Sources in Hydration and Recovery Drinks

Carbohydrate Type	Chemical Nature	Relative Sweetness (vs. Sucrose=100)	Osmolality Contribution	Absorption Rate	Key Benefits	Notes
Glucose (Dextrose)	Monosaccharide	70–75	High	Rapid	Immediate energy and hydration synerg	High osmolality; best below 6% concentration

					y with Na^+	
Sucrose	Disaccharide (glucose + fructose)	100	Moderate	Moderate	Balanced sweetness and stability	Common in isotonic drinks; natural claim-friendly
Fructose	Monosaccharide	120–130	Low	Slow	Smooth sweetness, lower glycemic index	Use ≤3% to avoid GI discomfort
Maltodextrin	Oligosaccharide (DE 10–20)	10–20	Low	Moderate	Adds body, low sweetness	Ideal for clear recovery drinks
Trehalose	Disaccharide	45	Low	Slow	Stabilizes proteins, gentle sweetness	Functional ingredient in endurance formulations

Isomaltulose (Palatinose™)	Disaccharide	40–50	Low	Slow	Sustained energy release; non-cariogenic	Premium ingredient for slow-carb beverages
Cluster Dextrin® (Highly Branched Cyclic Dextrin)	Polysaccharide	10	Very Low	Moderate–fast	Excellent GI tolerance, low osmolality	Advanced energy–hydration matrix

15.1.3 Carbohydrate Concentration and Osmotic Balance

Carbohydrate concentration directly affects osmolality, hydration efficiency, and gastric comfort:

Table 15.1.3: Carbohydrate–Osmolality Design Matrix - Target Ranges for Hypotonic, Isotonic, and Hypertonic Beverage Systems

Beverage Type	**Carbohydrate % (w/v)**	**Osmolality (mOsm/kg)**	**Function**
Hypotonic	1–4%	150–270	Rapid hydration, minimal energy
Isotonic	5–8%	280–320	Balanced hydration and energy delivery
Hypertonic	8–12%	320–400	Post-exercise recovery and glycogen replenishment

Guideline:

For endurance drinks, target 6–7% total carbohydrate, combining glucose (for fast uptake) and maltodextrin (for sustained release).
For electrolyte waters or light hydration drinks, stay below 3% total carbohydrate for optimal refreshment and clarity.

15.1.4 Synergy Between Carbohydrates and Electrolytes

Carbohydrates and electrolytes act synergistically:

- Sodium + glucose accelerate intestinal water absorption through co-transport.
- Potassium + fructose improve muscle glycogen storage post-exercise.
- Organic acid salts (citrate, lactate) reduce osmotic stress caused by high sugar loads.

This synergy underpins the success of the "dual-function isotonic matrix", typically 6% carbohydrate + 400–600 mg/L sodium for optimal hydration and energy balance.

15.1.5 Clean-Label and Low-Glycemic Carbohydrate Options

Modern consumers seek hydration beverages that deliver steady energy without sugar spikes or artificial ingredients.
Formulators can achieve this through smart carbohydrate selection:

Table 15.1.5: Smart Carbohydrate Strategies for Clean-Label Hydration - Glycemic Control, Fiber, and Natural Sweetness

Strategy	Ingredient Example	Effect
Reduced Sugar with Natural Sweeteners	Erythritol + Stevia	Maintains sweetness without calories
Low-Glycemic Slow-Release Energy	Isomaltulose, Trehalose	Sustained energy, no GI distress
Fiber-Enriched Hydration	Inulin (≤1%)	Adds prebiotic function, mild body

Natural Fruit Sugars	Coconut water solids, agave powder	Clean-label energy with minerals

15.1.6 Thermal and Storage Stability

Carbohydrates vary in their tolerance to processing and storage:

- Glucose and maltodextrin remain stable under hot-fill and UHT conditions.
- Sucrose may invert partially to glucose/fructose under acidic or thermal stress (pH <3.5, >85°C).
- Fructose darkens (Maillard reaction) in protein or amino acid systems, avoid in recovery drinks with whey or collagen.
- Isomaltulose and cluster dextrin resist browning and maintain clarity even in extended storage.

Processing Note: Always add carbohydrates before acids to prevent localized hydrolysis during blending.

15.1.7 Impact on Mouthfeel and Texture

Carbohydrates influence tactile perception as much as taste:

- Maltodextrin (0.2–0.6%) increases viscosity, giving a smooth "body" without excessive sweetness.
- Trehalose (0.1–0.3%) adds a soft, rounded texture suitable for recovery or relaxation beverages.
- Glycerol or polyols (when permitted) contribute additional smoothness and cooling perception.

Target Mouthfeel:

- Isotonic drinks → light, crisp finish (viscosity 1.2–1.5 cP)
- Recovery beverages → round, smooth mouthfeel (viscosity 1.6–2.0 cP)
- Functional waters → thin, ultra-clear texture (≤1.1 cP)

Formulation Insights

- Combine fast-acting glucose with slow-release maltodextrin or isomaltulose for sustained hydration energy curves.
- Keep total osmolality under 320 mOsm/kg to avoid gastric discomfort.
- For clean-label isotonic drinks, aim for 6% natural carbohydrate blend with citric–citrate buffer and pH 3.6–3.8.
- Adjust sweetness using blended systems (sucrose + stevia) to achieve 90–100% perceived sucrose sweetness with 30–40% calorie reduction.
- Carbohydrate design is central to sensory "roundness", fine-tune molecular weight distribution for both energy function and drinking satisfaction.

15.2 Designing Carbohydrate Blends for Performance, Recovery, and Clean-Label Applications

Single-source carbohydrates rarely meet the complex demands of hydration and energy beverages. A high-performance formula must supply immediate fuel, sustain energy over time, and support efficient rehydration, all while maintaining pleasant sweetness, clarity, and digestibility.

Designing carbohydrate systems for modern beverages therefore involves blending complementary sugars and polysaccharides to create controlled energy curves, balanced osmolality, and consumer-friendly labeling.

15.2.1 Principles of Carbohydrate Blending

Each carbohydrate contributes a unique **absorption rate**, **sweetness level**, and **metabolic impact**.
Blending them enables:

- Dual-phase energy release: rapid glucose uptake plus slower sustained supply.
- Optimized osmolality: combining low- and high-molecular-weight carbohydrates keeps total solute load within isotonic range.
- Improved taste and mouthfeel: adjusting sweetness and viscosity balance.

- Reduced GI distress: preventing "gut overload" from single rapid-absorption sugars.

Key Objective: Maintain total carbohydrate at 5–8% (isotonic), osmolality 280–320 mOsm/kg, and a glucose:fructose ratio near 2:1 for efficient utilization.

15.2.2 Dual- and Multi-Source Carbohydrate Systems

Table 15.2.2: Carbohydrate Blends and Performance Profiles

Blend Type	Components	Intended Function	Absorption Kinetics	Typical Use Level (% w/v)	Notes
Glucose + Maltodextrin	Fast + medium chain	Rapid + sustained energy	Rapid / Moderate	6–8	Classic isotonic blend; neutral flavor
Glucose + Fructose	Two intestinal transporters (SGLT1 + GLUT5)	Higher total carb uptake	Rapid / Moderate	5–7	Improves endurance tolerance
Maltodextrin + Isomaltulose	Low sweetness, sustained release	Smooth energy curve	Moderate / Slow	6–9	Excellent GI comfort; clean label
Glucose + Cluster Dextrin®	High-MW + rapid absorption	Long-duration hydration	Rapid / Slow	6–10	Premium endurance or recovery use

Sucrose + Maltodextrin + Trehalose	Balanced sweetness and stability	Energy + sensory harmony	Mixed	6–8	For flavored isotonic drinks
Isomaltulose + Erythritol (Low-Cal)	Functional low-GI system	Gentle energy; low glycemic load	Slow	4–6	Clean-label wellness hydration

Formulator's Note: Multi-source blends exploit multiple intestinal uptake pathways, increasing carbohydrate oxidation rate up to 1.6 g/min compared to ~1.0 g/min from glucose alone.

15.2.3 Matching Carbohydrate Profile to Performance Need

Table 15.2.3: Performance-Driven Carbohydrate Blending - Matching Energy Delivery Profiles to Activity Duration and Intensity

Application	Energy Requirement	Recommended Blend	Key Benefit
Pre-Workout Priming	Quick fuel availability	4% glucose + 2% maltodextrin	Rapid absorption without GI stress
During Endurance (1–3 hr)	Steady fuel + hydration	4% maltodextrin + 2% fructose	Maintains energy and isotonicity
Post-Workout Recovery	Glycogen restoration	4% glucose + 3% isomaltulose	Faster glycogen resynthesis
Functional / Low-Sugar Hydration	Minimal calories, taste structure	1–2% isomaltulose + polyol	Clean sweetness, stable pH

High-Performance Energy Drink	Sustained carb oxidation	5% maltodextrin + 1% cluster dextrin + 1% fructose	Controlled osmolality, long endurance support

15.2.4 Designing for Digestive Comfort and Performance Tolerance

High sugar concentration or single-path transport saturation often causes bloating or GI discomfort.
To minimize these effects:

- Maintain glucose + fructose ratio near 2:1 to use both SGLT1 and GLUT5 transporters.
- Keep osmolality ≤320 mOsm/kg.
- Avoid excess fructose (>3%) unless paired with glucose.
- Include electrolytes (Na^+ 400–600 mg/L) to facilitate co-transport and fluid uptake.
- Use low-DE maltodextrin to add body without sweetness or osmotic stress.

Practical Guideline:

Athletes tolerate blended carbohydrate solutions up to 90 g/hour of intake when multiple transporters are engaged.

15.2.5 Sweetness Management and Label Appeal

For clean-label and low-sugar claims, formulators can modulate perceived sweetness using natural synergy rather than synthetic sweeteners.

Table 15.2.5: Clean-Label Sweetness Architecture - Natural Modulation Strategies for Caloric Reduction and Label Appeal

Technique	Ingredient Example	Function

Sweetness Modulation	60% sucrose + 40% maltodextrin	Balanced taste with lower sugar load
Natural Sweetness Boosters	Stevia (Reb M), monk-fruit extract	Complements slow carbs without calories
Acid–Sweet Synergy	Citric/malic acid + isomaltulose	Perceived "brighter" sweetness
Mouthfeel Enhancers	Glycerol (0.2%) or inulin (≤1%)	Restores body lost by sugar reduction

Target Sweetness: Equivalent to 7–8% sucrose for isotonic drinks, 5–6% for wellness waters.

15.2.6 Engineering Osmolality and Stability

Carbohydrate molecular weight directly affects osmolality and clarity.

Table 15.2.6: Carbohydrate Molecular Weight vs. Osmolality Contribution - Design Implications for Clarity and Gastric Comfort

Carbohydrate	Avg. MW (Da)	Osmolality Contribution (per % w/v)	Comments
Glucose	180	High (~40 mOsm/kg)	Rapid absorption but higher osmotic load
Sucrose	342	Moderate (~25 mOsm/kg)	Stable and familiar
Maltodextrin (DE 10–20)	1800–3000	Low (~10 mOsm/kg)	Ideal for isotonic adjustment
Isomaltulose	342	Low (~15 mOsm/kg)	Sustained energy, mild sweetness
Cluster Dextrin®	~150 000	Very Low (~2 mOsm/kg)	Excellent clarity and endurance tolerance

Design Rule:

Use low-osmotic carriers (maltodextrin, cluster dextrin) to offset high-osmotic sugars (glucose, sucrose) and maintain clear, comfortable beverages.

15.2.7 Stability During Processing and Shelf Life

- Maintain pH 3.5–4.0 to avoid sucrose inversion and Maillard darkening.
- Avoid prolonged heat exposure (>90 °C, >1 min) in fructose-containing systems.
- Incorporate citric–citrate buffers to stabilize pH and taste during storage.
- For zero-preservative systems, use aseptic or cold-fill processing to prevent microbial sugar degradation.
- Add antioxidants (ascorbic acid, rosemary extract) when natural sugars are combined with fruit concentrates.

15.2.8 Example Formulation - Balanced Isotonic Energy Drink

Table 15.2.8: Balanced Isotonic Energy Drink Formulation

Ingredient	Function	% w/w
Purified water	Solvent	q.s. to 100
Glucose (anhydrous)	Rapid energy	3.0
Maltodextrin (DE 15)	Sustained energy / body	2.5
Fructose	Sweetness / secondary energy	1.0
Sodium citrate	Electrolyte + buffer	0.25
Potassium chloride	Electrolyte	0.10
Citric acid	Flavor balance / pH 3.6	0.12
Natural flavor	Sensory identity	0.02
Color (optional)	Visual cue	trace

Preservative (if required)	Shelf stability	0.02

Performance Parameters:

- Carbohydrate = 6.5%
- Osmolality ≈ 290 mOsm/kg
- pH = 3.6 ± 0.05
- Energy ≈ 26 kcal/100 mL
- Taste = moderate sweetness, crisp finish, no GI stress

Formulation Insights

- Combine fast and slow carbohydrates to tailor both taste and metabolic response.
- Maintain isotonic balance for comfort and absorption efficiency.
- For clean-label markets, pair maltodextrin + isomaltulose with natural flavors and acids for full functionality without "added sugar" stigma.
- Always measure real osmolality, not theoretical; deviations > ± 10 mOsm/kg can affect hydration kinetics.
- Taste balance drives adherence, energy without refreshment fails the user test.

15.3 Mouthfeel Engineering and Texture Modulation in Functional Beverages

Beyond flavor and electrolyte balance, mouthfeel defines how a beverage "feels alive" in the mouth, its smoothness, thickness, cooling, and lubricity. In hydration and nutrition drinks, achieving the right texture is both a sensory and functional engineering challenge: too thin feels watery and cheap, too thick hinders refreshment and gastric comfort.

Formulating the ideal mouthfeel requires careful control of viscosity, surface tension, and tactile persistence, achieved through carbohydrates, hydrocolloids, and natural texture enhancers.

15.3.1 The Science of Beverage Mouthfeel

Mouthfeel arises from the combined effects of:

- Viscosity: Resistance to flow, perceived as thickness or body.
- Surface tension: Influences smoothness and coating behavior on the tongue.
- Lubrication and cooling: Modified by polyols or glycerol.
- Rheological profile: Newtonian vs. pseudoplastic flow determines perception during sipping.
- Synergy with carbonation and acidity: Both sharpen tactile feedback, increasing perceived refreshment.

Target viscosity ranges (at 20°C):

Table 15.3.1: Target Viscosity Ranges for Hydration and Recovery Beverages by Functional Category

Beverage Type	Target Viscosity (cP)	Sensory Impression
Hypotonic / Electrolyte Water	1.0–1.2	Crisp, fast hydration
Isotonic Drink	1.2–1.5	Balanced body and flow
Recovery / Protein Beverage	1.6–2.2	Smooth, full mouthfeel
Functional Smoothie / Dairy-Based	2.5–3.5	Creamy, rich texture

15.3.2 Carbohydrates as Texture Agents

Carbohydrates are the primary mouthfeel builders in most hydration systems.

Table 15.3.2: Carbohydrates as Texture Agents - Mouthfeel Modulation and Osmotic Contribution

Ingredient	Function	Recommended Range (% w/w)	Notes
Maltodextrin (DE 10–20)	Adds viscosity and body without sweetness	0.2–0.6	Ideal for isotonic and recovery drinks
Trehalose	Softens acidity, smooths texture	0.1–0.3	Improves stability and taste longevity
Cluster Dextrin®	Enhances density and lubricity	0.5–1.0	High clarity and comfort, premium feel
Inulin (Soluble Fiber)	Adds mild creaminess, prebiotic claim	0.3–0.8	Use in low-pH or dairy-based systems
Erythritol / Glycerol	Cooling, lubricating mouthfeel	0.2–0.4	For functional or low-sugar waters

Formulator's Note: Pair maltodextrin and glycerol for isotonic beverages, the first provides body, the second smoothness.

15.3.3 Hydrocolloids and Natural Gums

Hydrocolloids act as structural stabilizers and subtle viscosity builders. Used at micro-levels, they prevent phase separation and impart silky flow without visible turbidity.

Table 15.3.3: Hydrocolloids for Mouthfeel and Stability - Functional Roles, Clarity Impact, and Usage Guidance

Hydrocolloid	Primary Effect	Typical Use (%)	Clarity	Notes

Xanthan Gum	Pseudoplastic body; stable at low pH	0.02–0.05	Clear	Provides smooth flow, stabilizes particulates
Gellan Gum (Low Acyl)	Light texture, high clarity	0.02–0.04	Excellent	Ideal for clear electrolyte and functional waters
Pectin (LM type)	Roundness, soft body	0.05–0.15	Slight haze	Works well with fruit acids
Guar Gum	Creamy texture	0.05–0.2	Opaque	Use in recovery or protein systems
Arabic Gum	Smooth mouthfeel, emulsion stability	0.1–0.3	Clear	Dual role as emulsifier and texturizer
Locust Bean Gum (LBG)	Thick, creamy feel	0.05–0.1	Opaque	Synergistic with guar for recovery drinks

Hydration Technique:

Hydrocolloids should be pre-dispersed in cold water under high shear, allowed to hydrate fully (≥30 min), then combined with acids and salts to avoid clumping or localized pH gelation.

15.3.4 Polyols and Natural Texture Modifiers

Polyols contribute both cooling sensation and viscosity modulation:

- Glycerol (E422): Enhances lubricity and slightly increases body; used at 0.2–0.5%.
- Erythritol: Adds gentle cooling and low-calorie bulk; use 0.3–0.6%.
- Sorbitol: Smoother texture but higher osmotic pressure, use with caution in isotonic systems.

- Betaine: Natural osmolyte; stabilizes mouthfeel under high salt conditions; functional and label-friendly.

Formulator's Tip: Combine glycerol (0.3%) with maltodextrin (0.3%) for a natural, cohesive "hydration texture" without sticky or heavy perception.

15.3.5 Engineering "Sensory Flow Curves"

Mouthfeel perception changes dynamically during sipping:

1. Initial Contact: Viscosity and cooling, controlled by maltodextrin and polyols.
2. Mid-Palate: Acidity, balance, and lubrication, influenced by pH, gums, and carbohydrate blend.
3. Afterfeel: Residual coating or dryness, managed by gum selection and electrolyte level.

Target profile: Quick onset (light), balanced mid (smooth), clean finish (non-coating).

Optimization Strategy:

- Use low-acyl gellan gum (0.02%) for quick-flowing isotonic drinks.
- Add 0.1% trehalose to round acidity and soften finish.
- For high-end functional beverages, a tri-blend (maltodextrin + glycerol + inulin) delivers balanced body and natural "hydration depth."

15.3.6 Texture–Flavor–Acidity Interactions

Texture modulates how flavor and acidity are perceived:

- Higher viscosity reduces perceived sourness and saltiness, useful for strong electrolyte systems.
- Mild gums increase aroma retention, prolonging perceived freshness.
- Polyols soften high-acid systems, preventing "bite."
- Thicker body enhances premium perception in low-sugar products.

Design Example:

A clear isotonic beverage (pH 3.6) with:

- Maltodextrin 0.3%
- Gellan gum 0.025%
- Glycerol 0.25%
  delivers 1.4 cP viscosity, bright citrus flavor, and smooth finish, ideal for athletes and wellness users alike.

15.3.7 Stability and Processing Considerations

- Always disperse gums under high shear before acid addition to avoid precipitation.
- Maintain pH >3.4 for optimal gum functionality and solubility.
- For cold-fill systems, ensure complete hydration (1 hour minimum) before final flavoring.
- Avoid overuse of gums in carbonated products, can destabilize CO_2 retention.
- Conduct shelf viscosity tracking (0, 1, 3, 6 months) to ensure no syneresis or thickening drift.

Quality Targets:

- Turbidity <1.0 NTU (clear beverages)
- Viscosity deviation ≤0.1 cP over 3 months
- No sedimentation, separation, or flavor dulling during storage

Formulation Insights

- Mouthfeel engineering transforms hydration from functional to experiential.
- Use minimal hydrocolloid loading for clarity, most benefits occur below 0.05%.
- Balance texture with flavor release: thicker systems mute aroma; lighter systems enhance volatility.
- Texture design can reinforce brand identity, “silky,” “clean,” or “rich” are reproducible sensory signatures.

- For low-calorie hydration waters, glycerol and low-acyl gellan gum offer a superior natural mouthfeel system with regulatory simplicity.

16. Protein Systems and Recovery Beverage Engineering

Protein-based hydration and nutrition beverages occupy a critical intersection between sports recovery, muscle maintenance, and functional nutrition. Beyond muscle repair, proteins also influence satiety, texture, and flavor perception, serving as both nutritional and sensory components.

The challenge in formulation lies in maintaining protein solubility, clarity, and flavor stability under acidic, heat-treated, or shelf-stable conditions. This chapter explores the biochemistry of protein performance, the technological hurdles of beverage integration, and strategies for optimizing stability and taste in both dairy and plant-based systems.

16.1 Functional Roles of Proteins in Post-Exercise and Nutritional Hydration Formulations

Proteins play a critical role in post-exercise and nutritional hydration formulations, serving functions that extend far beyond muscle repair. Their presence influences fluid retention, metabolic recovery, satiety, and the rebuilding of glycogen–protein complexes that support performance restoration. Different protein types and structures interact uniquely with acidic environments, electrolytes, and carbohydrate systems, shaping both physiological benefits and formulation behavior. Understanding these functional roles provides the foundation for designing hydration beverages that not only replenish fluids but actively accelerate recovery and rebuild performance capacity.

Following physical activity, the body requires three coordinated inputs for effective recovery:

1. Fluid and electrolyte replacement
2. Glycogen (carbohydrate) replenishment
3. Protein and amino acid resynthesis

The inclusion of proteins or peptides in a hydration beverage accelerates muscle protein synthesis (MPS) and enhances recovery kinetics, especially when co-ingested with carbohydrates.

16.1.1 Physiological Functions of Protein in Recovery Drinks

Table 16.1.1: Physiological Roles of Protein in Post-Exercise Recovery Beverages

Function	Biological Role	Formulation Implication
Muscle Repair	Supplies amino acids for myofibrillar protein synthesis	Determines protein type and peptide profile
Enhanced Glycogen Repletion	Stimulates insulin release when combined with carbohydrates	Supports carb-to-protein ratio of ~3:1
Rehydration Support	Improves plasma volume retention and osmotic balance	Requires soluble, non-coagulating proteins
Satiety and Energy Balance	Delays gastric emptying and provides sustained fullness	Relevant in nutrition and wellness drinks
Texture and Body	Adds natural viscosity and emulsification	Must be controlled for smooth mouthfeel

16.1.2 Optimal Carbohydrate–Protein Ratio

Research consistently supports the 3:1 to 4:1 carbohydrate-to-protein ratio for post-exercise recovery beverages.
This ratio:

- Replenishes glycogen more efficiently than carbohydrate alone.
- Enhances muscle repair through amino acid synergy.
- Reduces muscle soreness and fatigue during prolonged training.

Typical formula guideline:

- Carbohydrates: 6–8%

- Protein: 2–2.5% (whey isolate equivalent)
- Sodium: 400–600 mg/L
- pH: 3.8–4.2 (acidified milk-type) or 6.5–7.0 (neutral protein blend)

16.1.3 Protein Sources for Hydration and Recovery Beverages

Table 16.1.3: Common Protein Systems and Their Functional Attributes

Protein Source	Type	Solubility Range	Flavor Profile	Stability	Notes
Whey Protein Isolate (WPI)	Dairy	Excellent (pH 3–7)	Clean, light	High	Benchmark for sports recovery drinks
Whey Hydrolysate	Dairy (enzymatic hydrolysis)	Excellent	Slightly bitter	Very High	Fastest absorption; low allergenicity
Casein Hydrolysate	Dairy	Moderate	Mildly astringent	Medium	Sustained amino acid release
Pea Protein Isolate	Plant	Good (pH >6)	Earthy	Medium	Clean-label vegan alternative
Soy Protein Isolate	Plant	Good (pH 6–7)	Neutral–beany	Medium	Complete amino acid profile
Rice Protein	Plant	Moderate	Neutral–dry	Medium	Hypoallergenic; blend-friendly
Collagen Peptides	Animal	Excellent (pH 3–7)	Neutral	High	Supports joint and skin health; clear solution capable

Algal or Potato Protein	Plant	Moderate	Mild	Medium	Novel source with sustainable profile

16.1.4 Acidic vs. Neutral Protein Beverage Systems

Table 16.1.4: Acidic vs. Neutral Protein Beverage Systems

System Type	pH Range	Typical Protein	Processing Type	Stability Challenges
Acidic Protein Beverage	3.5–4.2	Whey isolate, collagen, hydrolysates	Hot-fill or aseptic	Protein denaturation, sedimentation
Neutral Protein Beverage	6.5–7.0	Casein, soy, pea blends	UHT / RTD (retort)	Aggregation, flavor oxidation

Key Difference:

At low pH, most native proteins unfold and lose solubility. Hydrolyzed or whey isolates resist this denaturation and are ideal for clear, shelf-stable protein hydration systems.

16.1.5 Enhancing Protein Solubility and Stability

a. pH and Buffer Control

- Maintain pH between 3.8–4.0 for clear whey systems; 6.7–6.9 for neutral recovery beverages.
- Use citric–citrate or phosphate buffers to prevent precipitation.

b. Ionic Strength Adjustment

- Add electrolytes after protein hydration to prevent flocculation.
- Avoid divalent ions (Ca^{2+}, Mg^{2+}) at high levels unless chelated (e.g., calcium citrate).

c. Hydrolysis and Particle Size

- Enzymatic hydrolysis improves solubility and heat stability.
- Target peptide molecular weight <5 kDa for clarity and absorption speed.

d. Temperature and Shear

- Hydrate proteins at 40–50°C under gentle agitation to prevent denaturation.
- Avoid prolonged high shear (>5000 rpm) once fully dispersed.

e. Chelators and Stabilizers

- Citric acid and sodium citrate help maintain ionic balance.
- Low-acyl gellan gum (0.02%) or pectin (0.1%) can stabilize acidified protein suspensions.

16.1.6 Example Formulation - Acidified Whey Protein Isolate Beverage

Table 16.1.6: Acid-Stable Whey Protein Isolate Beverage

Ingredient	Function	% w/w
Water	Solvent	q.s. to 100
Whey protein isolate	Functional protein	2.2
Sucrose or maltodextrin	Carbohydrate / sweetness	6.5
Citric acid	Acidifier / pH control	0.15
Sodium citrate	Buffer / electrolyte	0.25
Potassium chloride	Electrolyte	0.10
Natural flavor	Sensory balance	0.03
Ascorbic acid	Antioxidant	0.02
Color (optional)	Visual identity	trace

Key Parameters:

- pH = 3.8 ± 0.05
- Protein: 2.2%
- Carbohydrate: 6.5%
- Osmolality ≈ 300 mOsm/kg
- Appearance: clear, stable
- Thermal stability: >90°C for 30 sec (hot-fill compatible)

16.1.7 Flavor and Texture Considerations

- Protein hydrolysates often introduce bitterness or astringency, balance with fruit acids (malic, tartaric) and sweetener blends (stevia + erythritol).
- For smooth mouthfeel, combine collagen peptides (0.5%) + whey isolate (1.5%), adds body and stability.
- Acidified whey has a "light dairy note" that pairs best with citrus, berry, or tropical flavor bases.
- For plant proteins, mask earthy tones with caramelized, nutty, or spiced flavor layers.

Formulation Insights

- Use pre-hydrolyzed protein isolates for clarity and shelf-stable performance.
- Always pre-hydrate proteins before acid addition to prevent flocculation.
- Protein content above 2.5% requires stabilizer support (pectin, gellan).
- Post-blend filtration (0.5 μm) ensures clarity and prevents sedimentation.
- The most successful recovery beverages integrate protein + carbohydrate + electrolyte synergy within isotonic osmolality, creating functional, digestible systems with strong consumer acceptance.

16.2 Protein–Carbohydrate Synergy and Formulation Strategies for Recovery Systems

Hydration and energy restoration after exercise depend on three physiological pillars: water, electrolytes, and nutrient replenishment. While electrolytes restore plasma volume, and carbohydrates replenish glycogen, proteins accelerate muscle recovery and improve fluid retention through synergistic metabolic pathways.

Understanding and designing this synergy allows formulators to develop recovery drinks that not only rehydrate effectively but also promote cellular repair, glycogen synthesis, and sustained energy restoration.

16.2.1 Mechanisms of Protein–Carbohydrate Synergy

The body's post-exercise metabolism is uniquely sensitive to nutrient availability. Combining carbohydrates and proteins within 30–45 minutes post-activity triggers a cascade of beneficial responses:

Table 16.2.1: Mechanisms of Protein–Carbohydrate Synergy

Mechanism	Physiological Effect	Formulation Relevance
Insulin Co-Activation	Protein-derived amino acids (leucine, arginine) enhance insulin release beyond carbohydrate alone	Faster glucose uptake, improved glycogen restoration
Enhanced Glycogen Repletion	Co-ingestion accelerates glycogen storage by up to 38% vs. carbohydrate alone	Supports rapid recovery formulations
Protein Synthesis Activation	Amino acids, particularly BCAAs, activate mTOR pathway	Critical for muscle recovery and repair
Reduced Muscle Breakdown (Catabolism)	Maintains positive nitrogen balance	Essential for endurance athletes and high-intensity recovery

Fluid Retention and Plasma Expansion	Protein–electrolyte complexes improve osmotic water retention	Enhances hydration efficiency and thermoregulation

16.2.2 Ideal Nutrient Ratios for Recovery Formulations

Optimal macronutrient distribution for post-exercise drinks:

Table 16.2.2: Optimal Macronutrient and Physicochemical Targets for Post-Exercise Recovery Beverages

Component	Ideal Concentration	Function
Carbohydrate	5–8%	Glycogen replenishment and energy
Protein	1.5–3%	Muscle repair and recovery
Electrolytes (Na^+, K^+)	400–600 mg/L	Rehydration and muscle function
Osmolality	280–320 mOsm/kg	Efficient gastric emptying
pH	3.8–4.2 (acidic clear) or 6.6–6.8 (neutral)	Stability and flavor balance

Guideline Ratios:

- 3:1 carbohydrate-to-protein → Rapid recovery, endurance support.
- 2:1 carbohydrate-to-protein → Higher-protein variants for strength or repair-focused beverages.
- 4:1 carbohydrate-to-protein → Light recovery and isotonic hydration with protein enrichment.

16.2.3 Co-Transport and Absorption Dynamics

Co-ingestion enhances nutrient uptake by using overlapping absorption pathways:

- Glucose–Sodium Transport (SGLT1): Drives water and glucose absorption.
- Amino Acid–Sodium Transporters: Co-transport sodium and neutral amino acids, improving osmotic recovery.
- Insulin Pathway: Elevated insulin increases amino acid uptake into muscle tissue, accelerating repair.

This biochemical synergy means protein–carbohydrate beverages not only rehydrate but retain fluid more efficiently than carbohydrate-only drinks.

16.2.4 Designing Recovery Systems: Practical Matrix Selection

Table 16.2.4: Example Recovery Beverage Matrix Types

Product Type	Carbohydrate Source	Protein Source	Processing	Target pH	Key Benefits
Clear Protein Hydration Drink	Maltodextrin (3–4%) + glucose (2%)	Whey isolate (2%)	Hot-fill	3.8–4.0	Light, refreshing, shelf-stable
Creamy Recovery Beverage	Maltodextrin (5%) + sucrose (2%)	Casein hydrolysate (2%)	UHT	6.6–6.8	Sustained release, smooth texture
Plant-Based Recovery Drink	Isomaltulose (4%)	Pea + rice protein (2%)	Aseptic	6.8	Clean-label, vegan-friendly
Functional Collagen + Electrolyte Water	Fructose (2%) + maltodextrin (1%)	Collagen peptides (1.5%)	Cold-fill	3.6	Joint support, light mouthfeel

Endurance Formula (Extended Release)	Cluster dextrin (4%) + isomaltulose (3%)	Whey hydrolysate (1.5%)	Aseptic	4.0	Sustained energy + fast recovery

16.2.5 Managing Taste, Sweetness, and After feel

Protein–carbohydrate systems must balance taste (sweetness and acidity) and texture (body and smoothness).
Unmasked bitterness or dryness from proteins can undermine consumer acceptance.

Formulation Strategies:

- Use fruit-acid systems (citric + malic) to brighten flavor and suppress bitterness.
- Combine stevia (Reb M) with erythritol for clean sweetness and no metallic aftertaste.
- For neutral beverages, add vanilla, cocoa, or malt flavor bases to soften plant protein notes.
- Use inulin (0.3%) or glycerol (0.2%) to smooth mouthfeel and balance perception.

Example Sensory Profile Target (Recovery Beverage):

Table 16.2.5: Target Sensory Profile for Recovery Beverages - Quantitative Benchmarks for Sweetness, Acidity, and Mouthfeel

Attribute	Target Sensory Rating (1–5)
Sweetness	3.5
Acidity	3.0
Body / Smoothness	4.0
Aftertaste Cleanliness	4.5

Overall Balance	4.3

16.2.6 Stabilization and Processing Techniques

Protein–carbohydrate beverages are sensitive to heat, pH, and ion balance. To maintain clarity and shelf life:

- Order of Addition:
  Hydrate protein → adjust pH → add carbohydrate → blend electrolytes → adjust flavor → heat treat.
- Thermal Processing:
  - Acidic whey or collagen systems: 85–90°C for 30 sec (hot-fill).
  - Neutral plant/dairy systems: 135–140°C for 2–4 sec (UHT).
- Stabilizers:
  - Low-acyl gellan gum (0.02–0.04%) prevents sedimentation.
  - Pectin (0.05–0.15%) supports acidified whey or fruit-flavored systems.
- Filtration:
  0.5–1.0 µm filtration ensures clarity and removes denatured particulates.
- Storage:
  Store <30°C to prevent protein aggregation and Maillard darkening.

16.2.7 Example Formulation - Balanced Protein Recovery Beverage

Table 16.2.7: Balanced Post-Workout Recovery Beverage

Ingredient	**Function**	**% w/w**
Water	Solvent	q.s. to 100
Maltodextrin (DE 15)	Sustained energy	3.0
Glucose	Rapid energy	2.5

Whey protein isolate	Recovery protein	2.0
Citric acid	Acidity / flavor	0.10
Sodium citrate	Buffer / electrolyte	0.25
Potassium chloride	Electrolyte	0.10
Natural fruit flavor	Sensory	0.03
Ascorbic acid	Antioxidant	0.02
Sweetener (Stevia + Erythritol)	Sweetness balance	0.05
Color (optional)	Visual appeal	trace

Performance Parameters:

- Carbohydrate: 5.5%
- Protein: 2.0%
- Carbohydrate: Protein ratio = 2.75:1
- Osmolality ≈ 310 mOsm/kg
- pH = 3.9
- Shelf life: ≥9 months (ambient, hot-fill)
- Taste: Balanced fruit acidity with smooth, light texture

16.2.8 Emerging Trends: Beyond Basic Recovery

Modern recovery beverages are evolving beyond muscle restoration toward holistic performance and wellness.
Formulators now integrate:

- BCAA-enriched whey systems for rapid MPS activation.
- Collagen + electrolyte waters for joint and skin support.
- Plant protein + adaptogen blends (ashwagandha, maca) for stress recovery.

- Low-GI carbohydrate systems (isomaltulose + trehalose) for sustained energy release.
- Electrolyte + peptide synergies for rehydration and osmotic balance.

These next-generation systems represent the intersection of sports nutrition and functional wellness, blurring lines between hydration, recovery, and daily vitality.

Formulation Insights

- Optimal synergy occurs at a 3:1 carbohydrate: protein ratio with controlled osmolality (280–320 mOsm/kg).
- Hydrolyzed proteins enhance both nutrient absorption and processing stability.
- Maintain pH stability to prevent precipitation and Maillard discoloration.
- Protein–carbohydrate systems not only fuel performance, they extend brand versatility across recovery, endurance, and functional beverage lines.
- Consumer acceptance is sensory-driven: clarity, smoothness, and flavor masking determine repeat use more than nutritional label alone.

16.3 Protein Stability, Flavor Masking, and Processing Optimization in Acidic and Neutral Systems

Protein stability in beverages is one of the most technically demanding aspects of formulation.
Unlike carbohydrates and electrolytes, proteins are structurally fragile biopolymers that respond sensitively to pH, temperature, ionic strength, and processing shear. Inadequate stabilization can result in precipitation, sedimentation, phase separation, or off-flavors, all of which directly affect consumer perception and product shelf life.

This section provides the framework and best practices to engineer thermally, physically, and sensorially stable protein-based beverages, across both acidic and neutral pH systems.

16.3.1 Key Instability Mechanisms in Protein Beverages

Table 16.3.1: Key Instability Mechanisms in Protein Beverages and Mitigation Strategies

Instability Type	Cause	Typical Outcome	Mitigation Strategy
Thermal Denaturation	Protein unfolding under heat	Sediment, turbidity	Controlled pH, hydrolysis, pre-heating step
pH-Induced Aggregation	Protein near its isoelectric point (pI)	Flocculation, opacity	Formulate ≥0.5 pH units away from pI
Ionic Precipitation	Reaction with divalent cations (Ca^{2+}, Mg^{2+})	Sediment or haze	Chelate ions or use citrate buffers
Maillard Browning	Protein + sugar interaction under heat	Color change, flavor loss	Limit reducing sugars, control heat time
Oxidative Off-Flavors	Lipid or amino acid oxidation	Sulfuric or cardboard notes	Add antioxidants, oxygen scavengers
Phase Separation	Incomplete dispersion or emulsion instability	Layering, syneresis	Use hydrocolloids, homogenization control

16.3.2 Acidic Protein Beverage Systems (pH 3.5–4.2)

These are clear, light formulations often based on whey isolate, hydrolysates, or collagen peptides.
They are processed under hot-fill or cold-fill systems and rely on low ionic strength and small peptide size for stability.

Best Practices:

- Maintain pH 3.8–4.0, far from whey's isoelectric point (~5.2).
- Use hydrolyzed proteins (≤5 kDa) for heat tolerance.
- Pre-dissolve protein at neutral pH before acidifying to target.
- Avoid phosphate salts; prefer sodium citrate as a buffer.
- Use low-acyl gellan gum (0.02–0.04%) or pectin (0.1%) for stabilization.
- Pasteurize at 85–90°C for 30–45 seconds, or hot-fill at 88°C for 15 seconds.

Example:
A clear whey-based citrus drink (pH 3.9) with 2% protein and 6% carbohydrates remains stable for 12 months when buffered with sodium citrate and stabilized with 0.03% gellan gum.

16.3.3 Neutral Protein Beverage Systems (pH 6.6–7.0)

Neutral systems (casein, soy, pea, or mixed proteins) are typically milky or creamy and processed using UHT or retort sterilization for ambient shelf stability. Challenges include aggregation, heat denaturation, and flavor oxidation.

Stabilization Strategies:

- Pre-hydrate proteins at 50–55°C for full solubility.
- Maintain phosphate–citrate buffer (0.2–0.4%) for pH control.
- Add emulsifiers (lecithin, 0.1–0.2%) to prevent protein–fat aggregation.
- Use high-shear homogenization (150–200 bar) for uniform dispersion.
- Conduct UHT (135°C, 2–4 sec) or retort at 121°C for 15 min, depending on target shelf life.
- Include antioxidants (ascorbic acid, tocopherols) to minimize oxidation of sulfur amino acids and lipids.

Quality Monitoring:

- Protein solubility: ≥95% (by Kjeldahl nitrogen solubility index).

- Sedimentation: <1% by volume after 3 months at 40°C.
- Color stability: $\Delta E < 1.0$ after 12 weeks accelerated storage.

16.3.4 Flavor Masking and Sensory Balancing

Proteins, especially hydrolysates and plant isolates, introduce bitterness, astringency, and "green" or "beany" notes due to small peptides and residual lipids. These can be effectively managed through multi-layer flavor design:

Table 16.3.4: Sensory Masking and Flavor Architecture for Protein Beverages

Strategy	Ingredient / Technique	Sensory Effect
Acid–Sweet Balancing	Citric + malic acids with stevia	Brightens and rounds flavor
Masking Peptides	Vanilla, cocoa, malt, or caramel bases	Smooths bitterness
Flavor Layering	Combine citrus + floral or berry + cream	Distracts from off-notes
Polyol Cooling	Glycerol or erythritol (0.2–0.4%)	Refreshing, clean finish
Aroma Anchoring	Natural fruit esters or essential oils	Increases perceived freshness

Example:

A whey hydrolysate drink flavored with tropical citrus and sweetened with erythritol + stevia achieved 80% consumer liking in trials compared to 60% for the unflavored base.

16.3.5 Stabilizer and Texture System Design

Hydrocolloids can stabilize proteins both physically and texturally by preventing aggregation and improving mouthfeel.

Table 16.3.5: Hydrocolloids for Protein Beverage Stabilization

Stabilizer	pH Range	Function	Use Level (%)	Application Notes
Low-Acyl Gellan Gum	3.2–4.5	Protein suspension, clarity	0.02–0.04	Ideal for clear acid whey beverages
Pectin (LM type)	3.5–4.2	Protein–acid protection	0.05–0.15	Essential in fruit-flavored whey drinks
Carrageenan (κ)	6.5–7.0	Suspension and body in dairy systems	0.02–0.05	Synergistic with caseinates
Guar Gum	6.0–7.0	Creamy texture, viscosity	0.05–0.15	For plant protein milky systems
Arabic Gum	3.0–6.0	Emulsifier + texture balance	0.1–0.3	Enhances mouthfeel and flavor release

16.3.6 Processing Optimization and Order of Addition

Correct sequencing during blending is essential to avoid precipitation and uneven dispersion:

1. Dissolve protein in neutral water (pH 6.5–7.0).
2. Hydrate fully (30–45 min, 45°C) under gentle agitation.
3. Adjust pH downward gradually (if acidic system) using citric acid.
4. Add carbohydrates, then electrolytes, ensuring full dissolution.
5. Incorporate stabilizers under high shear.
6. Add flavors and sweeteners last.
7. Apply heat treatment (pasteurization, UHT, or hot-fill).
8. Homogenize (100–150 bar) post-heat if required for texture.

Pro Tip: For shelf-stable clarity, perform a fine filtration (0.2–0.5 μm) step before bottling.

16.3.7 Shelf-Life Extension and Quality Monitoring

- Storage Testing:
  Conduct accelerated aging at 40°C for 30 days and ambient for 6–12 months.
- Stability Markers:
  - pH drift ≤0.05 units
  - No visible sedimentation
  - Protein solubility ≥90%
  - No color or flavor degradation (ΔE <1.5)
- Packaging Considerations:
  - Use light-barrier PET or aluminum for protein-based drinks.
  - Nitrogen flushing reduces oxidative flavor loss.
  - Maintain oxygen <1 ppm at fill.

16.3.8 Example Processing Flow - Acid Whey Protein Beverage

1. Hydration: 2% whey isolate in water at 45°C, 30 min.
2. Acidification: Adjust to pH 3.9 with citric acid.
3. Buffering: Add sodium citrate and potassium chloride.
4. Stabilization: Incorporate 0.03% low-acyl gellan gum.
5. Sweetening: Add 6% maltodextrin and flavor system.
6. Homogenization: 120 bar single stage.
7. Hot-Fill: 88°C for 15 seconds, then cool to <30°C.
8. Storage: Shelf life 12 months at 25°C, clarity retained.

Formulation Insights

- Stability begins with solubility: pre-hydrate proteins before acid or salt exposure.
- pH and ionic control define clarity and long-term stability.
- Thermal mapping prevents denaturation without underprocessing.
- Flavor architecture should complement, not hide, protein notes.
- Protein beverages succeed when they feel clean, taste balanced, and stay stable for a full shelf life.

17. Functional Additives, Adaptogens, and Cognitive Hydration Enhancers

The modern beverage landscape has evolved far beyond basic hydration. Today's consumers expect drinks that restore, energize, calm, or focus, often simultaneously. The next generation of hydration systems integrates functional additives such as adaptogens, nootropics, botanical extracts, amino acids, and natural antioxidants to deliver measurable benefits for both body and mind.

This chapter provides a practical and scientific framework for designing such beverages, focusing on the physiological mechanisms, formulation compatibility, and stability strategies required to integrate these powerful bioactives into clear, shelf-stable, and sensory-balanced products.

17.1 The Evolving Role of Functional Additives in Next-Generation Hydration Systems

Functional additives turn hydration drinks into performance tools and wellness products. They can modulate stress response, improve focus, support recovery, and enhance mood. However, most of these bioactives are chemically sensitive and organoleptically challenging, requiring thoughtful formulation design to ensure efficacy, taste harmony, and shelf stability.

17.1.1 Functional Classification of Additives

Table 17.1.1: Functional Classification of Additives in Next-Generation Hydration Systems

Category	Key Compounds / Examples	Primary Benefits	Typical Dose Range	Solubility / Compatibility
Adaptogens	Ashwagandha, Rhodiola, Ginseng	Stress reduction,	100–300 mg/serving	Moderate; sensitive to heat

		cortisol modulation		
Nootropics	L-theanine, Caffeine, L-tyrosine	Cognitive focus, alertness, calm energy	50–200 mg/serving	High (except tyrosine, moderate)
Amino Acid Modulators	L-carnitine, Beta-alanine, Taurine	Fat metabolism, endurance, muscle recovery	250–1000 mg/serving	Excellent; pH-stable
Antioxidants	Vitamin C, Polyphenols, Green tea extract	Cellular protection, anti-fatigue	50–200 mg/serving	Variable; often stabilized in buffered systems
Electrolyte-Linked Botanicals	Coconut water solids, Sea minerals	Natural hydration and mineral complexity	0.5–2%	High; clear when filtered
Cognitive Hydration Actives	L-theanine + caffeine, Huperzine A	Mental clarity and hydration synergy	100–150 mg + 50 mg	pH-sensitive; protected forms preferred

17.1.2 Adaptogens: Supporting Stress Resilience and Recovery

Adaptogens are botanicals that help the body maintain homeostasis under stress, both physical and mental. Their integration into hydration beverages transforms them into functional wellness systems.

Table 17.1.2: Adaptogen Selection Matrix for Functional Hydration Systems

Adaptogen	Bioactive Compounds	Mechanism	Recommended Inclusion	Formulation Notes

Ashwagandha (With aniasomnifera)	Withanolides	Regulates cortisol, supports stamina	150–300 mg	Use aqueous extract; bitter, mask with cocoa, vanilla, or spice
Rhodiola rosea	Rosavin, salidroside	Enhances oxygen efficiency, mental focus	100–200 mg	Pair with berry or citrus flavor
Panax Ginseng	Ginsenosides	Boosts alertness, modulates fatigue	100–250 mg	Stable at low pH; combine with natural sweeteners
Cordyceps militaris	Cordycepin	Improves oxygen uptake, endurance	100–150 mg	Slight umami note; fits herbal profiles
Holy Basil (Tulsi)	Eugenol, ursolic acid	Stress relief, adaptogenic balance	100–200 mg	Pair with mint or lemongrass tones

Formulator's Note:

Adaptogenic actives are often hydrophobic and benefit from microencapsulation or emulsification (gum arabic, 0.2%) for clarity and bioavailability.

17.1.3 Nootropics and Cognitive Hydration Systems

Nootropics, cognitive enhancers, complement hydration and energy systems by improving mental clarity, focus, and calm energy without overstimulation. Their inclusion aligns with growing consumer demand for "mind + body" beverages that sustain performance without the crash of energy drinks.

Core Nootropic Pairings:

- L-theanine (100–150 mg) + Caffeine (50–75 mg) → calm alertness, smoother focus.
- L-tyrosine (200 mg) + B-vitamins (B6, B12) → neurotransmitter synthesis and mental stamina.
- Citicoline (CDP-Choline, 125 mg) → sustained cognitive energy and focus.

Formulation Considerations:

- L-theanine and caffeine are water-soluble and stable in pH 3–7.
- Tyrosine and choline derivatives require pH >4.5 and gentle thermal processing.
- Avoid combining high polyphenol extracts with amino nootropics, risk of precipitation or color drift.

Flavor Pairing Strategy:

- Citrus, mint, or tea-based profiles complement cognitive hydration systems.
- L-theanine + caffeine blends work well with lemon-lime or green tea citrus bases for a "clean energy" positioning.

17.1.4 Amino Acid-Based Functional Enhancers

L-carnitine, beta-alanine, taurine, and glutamine extend hydration into endurance and recovery. They function as metabolic modulators, improving fuel utilization and buffering fatigue.

Table 17.1.4: Amino Acid–Based Functional Enhancers - Roles, Dosing, and Integration Guidance

Amino Acid	Function	Typical Level	pH Stability	Formulation Note
L-Carnitine	Enhances fatty acid oxidation	250–500 mg	Excellent (3–7)	Slight sourness; complement with citrus

Beta-Alanine	Muscle carnosine precursor	500–1000 mg	Excellent	Combine with carbohydrate blend for uptake
Taurine	Osmoregulator and neural modulator	500–1000 mg	Excellent	Neutral flavor; enhances clarity
Glutamine	Recovery and immune support	250–500 mg	Moderate (degrades under high heat)	Use in cold-fill or aseptic systems

Formulator's Insight:

Taurine can also serve as a functional osmolyte, improving cellular water retention, ideal for advanced rehydration systems.

17.1.5 Botanical Antioxidants and Polyphenols

Polyphenols from plants offer antioxidant protection, color depth, and flavor complexity, but they present formulation challenges such as instability, astringency, and oxidative browning.

Table 17.1.5: Botanical Extracts for Antioxidants and Polyphenols

Extract	Key Compounds	Benefits	pH/Heat Sensitivity	Recommended Strategies
Green Tea Extract	Catechins	Antioxidant, metabolism support	High	Stabilize with ascorbic acid (0.02%)
Grape Seed Extract	OPCs	Vascular health, anti-fatigue	Medium	Use encapsulated forms to prevent color drift

Turmeric (Curcumin)	Curcuminoids	Anti-inflammatory	High	Nanoemulsion (0.5–1%) for solubility
Elderberry / Acai	Anthocyanins	Immune and antioxidant support	High	Use stabilized anthocyanin colors
Matcha / Yerba Maté	Polyphenols, caffeine	Alertness and antioxidant	Medium	Use cold-fill to preserve flavor and color

Flavor Balancing Tip:

Blend polyphenol-rich extracts with malic acid (0.05%) to lift brightness and erythritol (0.2–0.4%) to round astringency.

17.1.6 Integration Guidelines for Functional Additives

Table 17.1.6: Integration Protocol for Functional Additives in Hydration Beverages

Parameter	**Guideline**
Order of Addition	Add extracts and actives post-dissolution of salts and acids.
Thermal Processing	For polyphenols or adaptogens, use ≤85°C for <1 min (hot-fill) or cold-fill aseptic.
pH Stability Window	Maintain between 3.6–4.2 for most actives and acids.
Clarity Management	Use gum arabic or low-acyl gellan gum at 0.02–0.05% to prevent haze.
Microencapsulation	For sensitive actives (curcumin, caffeine, herbal oils), encapsulate in maltodextrin or starch systems.

Oxygen Protection	Include 50–100 ppm ascorbic acid or tocopherols; nitrogen flush before capping.

17.1.7 Example Formulation - Adaptogenic Cognitive Hydration Drink

Table 17.1.7: Adaptogenic Cognitive Hydration Drink

Ingredient	Function	% w/w
Water	Solvent	q.s. to 100
Glucose	Hydration energy	2.0
Isomaltulose	Sustained energy	2.0
Sodium citrate	Electrolyte + buffer	0.25
Potassium chloride	Electrolyte	0.10
Ashwagandha extract (aqueous)	Adaptogen	0.03
L-theanine	Cognitive focus	0.015
Caffeine (natural green tea source)	Energy	0.007
Citric acid	pH 3.8 adjustment	0.10
Natural flavor (citrus–mint)	Sensory	0.03
Stevia (Reb M)	Sweetness	0.02
Ascorbic acid	Antioxidant protection	0.02
Color (optional)	Visual identity	trace

Performance Metrics:

- pH: 3.8
- Osmolality: ~285 mOsm/kg
- Clarity: <1.0 NTU

- Energy: ~18 kcal/100 mL
- Cognitive Support: Balanced caffeine/theanine profile for calm alertness
- Stability: 9–12 months (hot-fill or aseptic)

Formulation Insight

- Functional hydration is no longer binary (thirst or energy); it's multi-dimensional, addressing stress, focus, and endurance together.
- Actives must be protected, solubilized, and sensory-harmonized, clarity and taste remain as vital as efficacy.
- Balance adaptogens (calm) with nootropics (focus) and electrolytes (hydration) to create holistic performance systems.
- Functional additives should enhance physiological resilience while supporting modern sensory expectations: clean, crisp, and naturally flavored.

17.2 Adaptogen and Nootropic Synergy - Designing Mind–Body Hydration Systems

Modern functional hydration is moving beyond replenishment into cognitive balance and emotional wellness.
The fusion of adaptogens (which normalize stress response) and nootropics (which optimize cognitive function) allows beverage formulators to engineer mental–physical synergy: drinks that energize the body while calming the mind.

This section outlines the physiological rationale, design principles, and formulation strategies for these new-generation "mind–body" hydration systems.

17.2.1 Physiological Rationale: Balancing Cortisol and Neurotransmitters

Physical exertion and daily stress both elevate cortisol and adrenaline, which deplete hydration reserves and impair focus.
Adaptogens such as ashwagandha or rhodiola moderate this hormonal stress

cascade, while nootropics like L-theanine and tyrosine fine-tune neurotransmitters including dopamine, GABA, and serotonin.

When combined, these actives produce dual-phase cognitive hydration:

- Phase 1: Calm alertness, L-theanine + caffeine smooth stimulation, preventing "jitter."
- Phase 2: Adaptive resilience, Ashwagandha or Rhodiola stabilize mood, lower stress response.
- Phase 3: Sustained focus, Tyrosine and B-vitamins support neurotransmitter synthesis and mental stamina.

Outcome: A beverage that restores hydration, enhances mental performance, and buffers the physiological impact of stress.

17.2.2 Functional Pairing Strategies

Table 17.2.2: Adaptogen–Nootropic Pairing Matrix

Functional Goal	Adaptogen	Nootropic / Amino Acid	Supporting Nutrients	Formulation Notes
Calm Focus	Ashwagandha (200 mg)	L-theanine (100 mg) + low caffeine (50 mg)	Magnesium, B6	Excellent for daily cognitive hydration
Mental Clarity	Rhodiola (150 mg)	L-tyrosine (200 mg)	Citicoline, B12	pH >4.0; best cold-fill
Stress Relief + Relaxation	Holy Basil (150 mg)	GABA (100 mg) or L-glycine (200 mg)	Zinc, inositol	Nighttime or adaptogenic water line
Alert Energy	Ginseng (200 mg) + Rhodiola (100 mg)	Caffeine (75 mg) + L-	Electrolytes, B-complex	For focus-oriented hydration or

		theanine (100 mg)		productivity drinks
Mood and Cognitive Balance	Reishi (100 mg) + Ashwagandha (150 mg)	L-serine, taurine	Vitamin C, magnesium	Calming, immune-support beverage type

Formulator's Insight:

Synergistic effects arise when neural stimulants and adaptogens are balanced at physiological rather than pharmacological doses, the key to maintaining drinkability and compliance.

17.2.3 Ingredient Compatibility and Solubility Matrix

Table 17.2.3: Solubility and Stability Profile of Cognitive Actives

Compound	Solubility	pH Range	Heat Stability	Notes
L-Theanine	Excellent	3–7	High	Neutral flavor, stable under hot-fill
L-Tyrosine	Moderate	>4.5	Moderate	Pre-dissolve at warm temperature (50°C)
Caffeine (Natural / Anhydrous)	Excellent	3–7	High	Use natural green tea source for clean label
Ashwagandha (Aqueous Extract)	Moderate	3.5–6.5	Low	Use encapsulated or spray-dried powder for clarity
Rhodiola Extract	Moderate	3.8–6.5	Medium	Pair with berry flavor; mask herbal notes
Citicoline (CDP-Choline)	Excellent	4–7	High	Ideal for clear focus beverages

GABA	Excellent	3–7	High	pH-stable and neutral-tasting

Stability Optimization:

- Maintain pH 3.8–4.0 for most blends.
- Avoid prolonged heat exposure for botanical actives; prefer cold-fill or short hot-fill (<85°C, 30 sec).
- Use gum arabic (0.1–0.2%) as a carrier to solubilize hydrophobic adaptogens.
- Incorporate citric acid + sodium citrate buffer to stabilize pH and prevent haze.

17.2.4 Cognitive Hydration Matrix Design

Designing "smart hydration" beverages involves aligning mental performance outcomes with hydration dynamics.

Core Design Elements:

1. Electrolytes (Na^+, K^+, Mg^{2+}) → sustain hydration and neurotransmitter conductivity.
2. Carbohydrate base (2–3%) → mild energy support without glycemic spikes.
3. Caffeine–Theanine Ratio 1:2 → balanced energy and calm focus.
4. Adaptogenic backbone (Ashwagandha / Rhodiola) → resilience against stress fatigue.
5. B-complex and Vitamin C → neurotransmitter synthesis and oxidative protection.
6. Citric–malic acid system → crisp taste, controlled pH (~3.8).

17.2.5 Example Formulation - Cognitive Adaptogenic Hydration Drink

Table 17.2.5: Cognitive Adaptogenic Hydration Beverage

Ingredient	Function	% w/w

Purified Water	Solvent	q.s. to 100
Glucose	Energy / co-transport	2.0
Sodium Citrate	Buffer + electrolyte	0.25
Potassium Chloride	Electrolyte	0.10
L-Theanine	Calm focus	0.015
Caffeine (natural source)	Mental energy	0.007
Ashwagandha extract (encapsulated)	Adaptogenic balance	0.03
Citicoline (CDP-Choline)	Cognitive function	0.012
Citric acid	Acidity control (pH 3.8)	0.10
Natural flavor (citrus–mint–green tea)	Sensory	0.03
Stevia (Reb M)	Sweetness	0.02
Ascorbic acid	Antioxidant	0.02
Gum Arabic	Solubility carrier	0.05

Specifications:

- pH: 3.8 ± 0.05
- Osmolality: ~280 mOsm/kg
- Energy: ~15 kcal/100 mL
- Clarity: <1.0 NTU
- Stability: ≥9 months (cold-fill)
- Taste: Bright, refreshing, calm-focus profile

17.2.6 Sensory Balancing and Flavor Strategy

Functional hydration drinks often struggle with herbal or bitter notes.
A successful flavor profile integrates sensory masking + psychological signaling (flavors associated with calm, energy, or clarity).

Table 17.2.6: Sensory Architecture for Mind–Body Beverages

Effect Desired	Flavor Archetype	Example Components
Calm / Relaxation	Mint, cucumber, chamomile	Ideal for L-theanine + ashwagandha systems
Focus / Clarity	Lemon-lime, green tea, citrus zest	Matches caffeine–citicoline blends
Mood Boost / Refreshment	Berry, hibiscus, tropical fruit	Masks rhodiola or ginseng notes
Nighttime Relaxation	Vanilla, lavender, honey	Supports GABA- or glycine-based drinks

Formulation Tip:

Use dual-acid systems (citric + malic) to increase perceived brightness while minimizing bitterness.
For herbal extracts, include 0.05–0.1% glycerol to smooth aftertaste.

17.2.7 Labeling and Regulatory Considerations

- All adaptogens and nootropics must comply with local nutraceutical and food-supplement regulations.
- Clearly indicate functional claims as "supports focus," "reduces fatigue," or "promotes calm energy," avoiding medicinal wording.
- Use standardized extracts with declared active percentages (e.g., Ashwagandha root extract, 5% withanolides).
- Ensure daily intake levels do not exceed regional upper limits (e.g., caffeine <200 mg/day for general use).
- Clean-label trend: prefer botanical extracts and natural amino acids over synthetic nootropics for better market acceptance.

Formulation Insight

- Mind–body hydration systems succeed when stimulus and calm coexist, too much caffeine undermines the adaptogenic effect.
- Balance functional doses with taste architecture; most users prefer perceptible freshness rather than strong herbal flavor.
- Pair polyphenol antioxidants (green tea, citrus bioflavonoids) with adaptogens for a "wellness plus performance" signature.
- The future of hydration is holistic, restoring mental clarity, emotional stability, and cellular hydration in one synergistic system.

18. Natural Colorants, Antioxidants, and Phytonutrient Systems

Color is the first sensory signal of a beverage's identity. A bright red electrolyte drink suggests energy and vitality; a soft green or amber tone implies calmness and balance.
In functional hydration systems, color is more than aesthetic, it reflects chemical stability, natural sourcing, and antioxidant potency.

Natural pigments, polyphenols, and carotenoids can also act as bioactive phytonutrients, contributing antioxidant and anti-inflammatory properties that enhance recovery and cellular protection.
However, their integration presents unique challenges: most natural colorants are pH-, light-, and heat-sensitive, requiring precise formulation control.

This chapter outlines practical approaches for formulating beverages with stable, natural color systems that align with clean-label, functional, and sensory expectations.

18.1 Functional Roles of Natural Pigments and Antioxidant Components in Beverage Design

Natural pigments and antioxidant components bring dual functionality to beverage design, shaping both the visual identity and the biochemical resilience of a formulation. These compounds provide vibrant, consumer-driven color while simultaneously protecting flavors, nutrients, and delicate aroma systems from oxidative degradation. Their behavior is influenced by pH, light exposure, electrolyte load, and processing conditions, making their integration a strategic decision rather than a purely aesthetic one. Understanding their functional roles enables developers to create beverages that are visually compelling, nutritionally reinforced, and stable across real-world shelf-life challenges.

18.1.1 The Dual Role of Natural Colorants

Natural colorants serve two overlapping purposes:

1. Sensory and Branding Role:
   - Defines category expectations (e.g., citrus = yellow, berry = red).
   - Enhances perceived freshness and flavor intensity.
   - Serves as a visual cue for function, energizing, calming, or detoxifying.
2. Functional and Nutritional Role:
   - Many pigments (anthocyanins, carotenoids, chlorophylls) are antioxidant molecules themselves.
   - Scavenge free radicals, protect cells from oxidative stress.
   - Stabilize other ingredients (vitamins, flavor oils) through redox buffering.
   - Reinforce "natural origin" and "clean label" claims.

18.1.2 Classification of Natural Colorants and Their Stability Behavior

Table 18.1.2: Natural Colorant Classification

Colorant Class	Natural Source	Typical Hue	pH Stability Range	Light & Heat Stability	Key Functional Benefit
Anthocyanins	Berries, red cabbage, hibiscus	Red–purple	2.5–4.0	Moderate	Antioxidant, immune support
Betalains	Beetroot, amaranth	Red–violet	3.5–6.0	Low	Anti-inflammatory, blood flow support
Carotenoids	Carrot, paprika, turmeric	Yellow–orange	3.0–7.0	High (encapsulated)	Antioxidant, vitamin A precursor

Chlorophylls / Chlorophyllins	Spinach, alfalfa	Green	6.0–8.0	Low	Detox and alkalinity perception
Curcuminoids	Turmeric	Bright yellow	3.0–7.0	Moderate	Anti-inflammatory, antioxidant
Phycocyanins	Spirulina extract	Blue	4.0–7.0	Low–moderate	Cellular protection, oxygen metabolism
Caramel & Maillard pigments	Thermal sugars	Amber–brown	2.5–7.0	Very High	Flavor reinforcement, warmth

18.1.3 Antioxidants and Phytonutrients in Beverage Formulation

Antioxidants are often added to functional hydration systems for three interlinked reasons:

1. Protection of color and flavor, prevents oxidation of natural pigments and aromas.
2. Physiological benefit, supports cellular defense and recovery.
3. Shelf-life extension, minimizes off-flavor formation during storage.

Table 18.1.3: Key Natural Antioxidants and Their Roles

Compound	**Natural Source**	**Function**	**Effective Dose (ppm)**	**Stability Note**

Ascorbic acid (Vitamin C)	Citrus, acerola	Reduces oxidative degradation	100–300	Stable at low pH
Tocopherols (Vitamin E)	Plant oils	Lipid-phase antioxidant	50–200	Protects emulsified flavor oils
Polyphenols (Catechins, OPCs)	Tea, grape seed	Antioxidant + vascular support	100–300	Sensitive to light and oxygen
Rosemary extract	Herb	Natural preservative / antioxidant	50–150	Oil-soluble, mild herbal note
Green coffee extract	Coffee fruit	Antioxidant + mild stimulant	100–200	Stable at pH >4.0
Curcumin	Turmeric	Antioxidant, anti-inflammatory	50–100	Use nano emulsified form for clarity

Formulation Tip:

Combine ascorbic acid (water phase) and tocopherols (oil phase) for full-spectrum protection, this dual system controls oxidation in both aqueous and flavor components.

18.1.4 Designing Color–Function Relationships

Color can act as a functional language in beverage design, linking visual identity to physiological benefit.

Table 18.1.4: Color–Function Mapping Framework - Chromatic Identity, Ingredient Basis, and Physiological Signaling in Functional Hydration Systems

Hue Range	Typical Ingredient Base	Associated Functional Message
Bright Yellow / Orange	Turmeric, carotenoids, citric acids	Energy, vitality, metabolism
Red / Purple	Anthocyanins, betalains	Recovery, antioxidant, immunity
Blue / Green	Spirulina, chlorophyll	Calm, focus, detox
Amber / Brown	Herbal extracts, caramel	Comfort, grounding, relaxation
Translucent / Clear	Mineral, coconut, adaptogen systems	Purity, hydration, mental clarity

Branding Insight:

Align color psychology with beverage purpose, for instance, a "calm focus" drink benefits from pale green or blue hues, while a "performance hydration" drink signals efficacy with vivid yellow or orange.

18.1.5 Stabilizing Natural Colors in Electrolyte and Acidic Systems

Natural colors are especially unstable in acidic, ionic environments. Formulation success relies on pH control, chelation, and antioxidant co-factors.

Stability Enhancement Techniques:

- Buffer Control: Maintain pH within pigment stability window (anthocyanins: 3.0–3.8).
- Chelation: Use citric acid or EDTA to bind metal ions that catalyze oxidation.
- Oxygen Exclusion: Nitrogen flush at filling; maintain dissolved O_2 <1 ppm.
- Light Protection: Use amber PET, shrink sleeves, or aluminum cans.

- Antioxidant Pairing: Combine colorant with ascorbic acid + tocopherol for synergy.
- Encapsulation: Use maltodextrin or gum arabic (0.5–1%) to shield pigments.
- Thermal Control: Limit exposure >85°C to <30 sec; prefer aseptic cold-fill for delicate pigments.

Example:

A hibiscus-based electrolyte drink (pH 3.6, 0.05% anthocyanin) stabilized with ascorbic acid (0.03%) and citric buffer retained >90% color intensity after 6 months ambient storage.

18.1.6 Example Formulation - Natural Antioxidant Hydration Drink

Table 18.1.6: Natural Antioxidant Hydration Beverage Formulation

Ingredient	**Function**	**% w/w**
Purified water	Solvent	q.s. to 100
Isomaltulose	Mild energy base	2.0
Sodium citrate	Buffer + electrolyte	0.25
Potassium chloride	Electrolyte	0.10
Citric acid	Acid balance (pH 3.7)	0.12
Hibiscus extract	Natural color + polyphenols	0.05
Ascorbic acid	Antioxidant	0.03
Rosemary extract (water dispersible)	Secondary antioxidant	0.01
Stevia (Reb M)	Sweetener	0.02

Natural flavor (berry–citrus)	Sensory balance	0.03
Color adjustment (natural, optional)	Branding tone	trace

Performance Characteristics:

- pH: 3.7
- Osmolality: 280 mOsm/kg
- Color: Deep ruby red (anthocyanin stabilized)
- Clarity: <1.2 NTU
- Shelf life: ≥9 months (ambient, protected from light)
- Sensory: Bright, crisp, antioxidant-rich flavor profile

Formulation Insight

- Natural colorants are both visual and bioactive ingredients, treat them as functional nutrients, not just additives.
- Always match pigment chemistry with beverage pH and ionic strength.
- Antioxidant systems extend both color life and perceived freshness.
- Use encapsulation or buffered systems to achieve clear, shelf-stable, natural colors in functional hydration lines.
- Color consistency across batches reinforces brand trust and perceived product quality.

18.2 Stabilization Strategies for Natural Colorants, Polyphenols, and Carotenoids in Functional Beverages

Natural pigments and phytonutrients are among the most valuable, and challenging, components in modern beverage formulation.
They provide visual appeal and functional benefits but are chemically fragile, reacting to light, oxygen, pH shifts, and ions present in electrolyte systems.

This section outlines advanced stabilization methods that preserve the intensity, clarity, and bioactivity of natural colorants and antioxidants, ensuring consistency through production, storage, and distribution.

18.2.1 The Chemistry of Instability

Color and antioxidant degradation often stem from:

- Oxidation (exposure to oxygen and metal ions)
- Photodegradation (UV or visible light exposure)
- pH-dependent tautomeric shifts (anthocyanin color loss)
- Thermal breakdown (heat-sensitive carotenoids and betalains)
- Polyphenol–protein complexation (haze and sedimentation)

These reactions not only reduce visual vibrancy but can also generate off-flavors, turbidity, and loss of health claim validity (due to active degradation).

18.2.2 Anthocyanin Stability Enhancement

Anthocyanins are the most common natural red–purple pigments, yet also the most pH-sensitive.
At pH below 3.5, they appear red; at higher pH, they shift toward blue or lose color intensity.

Stabilization Strategies:

Table 18.2.2: Anthocyanin Stabilization Matrix

Technique	Mechanism	Typical Usage	Notes
pH Control (3.0–3.8)	Maintains red flavylium cation form	Citric or malic acid buffer	Ideal for electrolyte beverages
Co-pigmentation	Flavonoid stacking increases color density	Add quercetin or rutin (trace)	Natural enhancement effect

Antioxidant Coupling	Protects from oxidation	Ascorbic acid 0.02–0.05%	Dual benefit for shelf life
Chelation	Binds metal ions that catalyze oxidation	EDTA (if permitted) or citric acid	Prevents color fading
Encapsulation	Shields anthocyanins from light and oxygen	Maltodextrin, acacia gum	For clear beverages
Temperature Control	Minimize heat exposure	≤85°C for <30 sec	Cold-fill preferred

Formulator's Note:

Anthocyanins from black carrot, purple sweet potato, or hibiscus show superior heat and pH stability compared to blueberry or grape-derived pigments.

18.2.3 Betalains: Vibrant Yet Heat-Sensitive

Betalains deliver vivid red-violet hues from natural sources like beetroot but degrade under prolonged heating and oxygen exposure.

Best Practices for Betalain Systems:

- Maintain pH 3.5–5.5, avoid strong acids.
- Add ascorbic acid (0.03–0.05%) to reduce oxidative browning.
- Use nitrogen sparging before filling to remove dissolved oxygen.
- Prefer cold-fill or aseptic processing.
- Incorporate polysaccharides (gum arabic, pectin 0.1%) to enhance pigment retention and prevent sedimentation.

Example:

A beet–apple electrolyte drink retained >90% color intensity after 6 months when packed in light-block PET and stored below 25°C.

18.2.4 Carotenoids: Lipid-Soluble Color Stability

Carotenoids such as β-carotene, lutein, and lycopene are powerful antioxidants but insoluble in water.
They require emulsification or microencapsulation to disperse uniformly in beverages.

Stabilization Strategies for Carotenoid Systems:

Table 18.2.4: Carotenoid Delivery and Stability System

Technique	Function	Ingredient Example
Nanoemulsification (100–200 nm)	Enhances optical clarity and absorption	Lecithin, polysorbate 80
Natural Emulsifiers	Improves clean label compatibility	Quillaja saponins, gum arabic
Antioxidant System	Protects from oxidation	Tocopherol + ascorbyl palmitate
Thermal Control	Limits degradation	<90°C short time exposure
Packaging	Light barrier required	Amber PET, aluminum can

Formulation Insight:

Use turmeric-derived curcumin emulsions (0.01–0.05%) for bright golden-yellow hues with antioxidant function; stable at pH 3–5.

18.2.5 Chlorophyll and Phycocyanin: Green and Blue Challenges

Chlorophylls and spirulina-derived phycocyanins offer unique natural colors but are highly sensitive to acid and light.

Stabilization Guidelines:

- Maintain pH above 4.5 for green shades; below 3.5 causes browning or fading.
- For acid systems, use copper chlorophyllin, a more stable derivative.
- Keep away from light; store in opaque containers.
- Avoid high ionic strength, excessive electrolytes accelerate pigment breakdown.
- Cold-fill recommended; hot-fill can irreversibly damage protein-based pigments like phycocyanin.

Natural Blue Options:

Use spirulina extract stabilized with trehalose (0.2%) or blue gardenia extract for improved color retention in pH 4–5 beverages.

18.2.6 Polyphenol and Antioxidant Co-Stabilization

Polyphenols (catechins, flavonoids) act as antioxidants but oxidize rapidly in open systems, leading to browning or astringency.

Stability Optimization Steps:

- Use chelating acids (citric or tartaric) to prevent metal-induced oxidation.
- Add secondary antioxidants (rosemary extract, tocopherol) to slow radical propagation.
- Maintain low dissolved oxygen (<1 ppm) during mixing and filling.
- Include natural sweeteners (erythritol, honey solids) to mask bitterness and enhance mouthfeel.
- Protect from UV light using opaque or shrink-sleeve packaging.

Example:

Green tea polyphenols retained >85% catechin potency for 9 months when formulated with 0.02% ascorbic acid, 0.1% sodium citrate, and amber PET packaging.

18.2.7 Advanced Stabilization Technologies

Table 18.2.7: Advanced Stabilization Technologies for Natural Colorants and Bioactive Pigments

Technology	Mechanism	Application	Benefit
Spray-Dried Encapsulation	Entraps pigment in carbohydrate matrix	Anthocyanins, carotenoids	Excellent dry-blend stability
Liposomal Encapsulation	Phospholipid vesicles protect actives	Polyphenols, vitamins	Enhances bioavailability
Microbead Coating (Alginate, Gelatin)	Physical isolation	Natural pigment blends	Controlled release and protection
Antioxidant Layering	Sequential addition of antioxidants	Complex functional systems	Delays oxidative chain reaction
Electrostatic Emulsions	Stabilizes lipid-soluble colorants	Carotenoid beverages	Clear, shelf-stable emulsions

18.2.8 Example Formulation - Stable Natural Color + Antioxidant Beverage Base

Table 18.2.8: Stable Natural Color–Antioxidant Beverage Base

Ingredient	Function	% w/w
Purified water	Solvent	q.s. to 100
Citric acid	pH control (3.8)	0.10
Sodium citrate	Buffer + electrolyte	0.25
Ascorbic acid	Antioxidant	0.03

Natural colorant (black carrot anthocyanin)	Pigment + phytonutrient	0.05
Tocopherol (Vit. E emulsion)	Lipid-phase antioxidant	0.02
Gum Arabic	Emulsifier / stabilizer	0.05
Potassium chloride	Electrolyte	0.10
Natural flavor (berry–citrus)	Sensory	0.03
Stevia (Reb M)	Sweetness	0.02

Specifications:

- pH: 3.8
- Clarity: <1.0 NTU
- Color Retention: ≥90% (6 months ambient)
- Bioactive Retention: ≥85% polyphenol potency
- Recommended process: cold-fill, nitrogen-flushed, light-protected packaging

18.2.9 Formulation Insight

- Always treat colorants as sensitive actives, not mere aesthetic additives.
- Integrate antioxidant systems and encapsulation early in formulation.
- The ideal beverage color system combines chemical stability, regulatory safety, and functional alignment with brand positioning.
- Future trends favor bioactive pigment systems, color ingredients that also contribute to health, such as curcumin for inflammation or anthocyanins for vascular health.

18.3 Co-Formulation of Natural Antioxidants and Electrolyte Systems for Enhanced Stability and Functional Performance

The combination of antioxidants and electrolytes defines the heart of next-generation hydration beverages: electrolytes restore physiological balance, while antioxidants protect the body, and the product, from oxidative stress.
However, bringing these two systems together creates formulation challenges because the ionic strength and metal ions in electrolyte salts can accelerate oxidative degradation of sensitive colorants, polyphenols, and vitamins.

This section details strategies to harmonize these two functional systems so they complement, rather than compromise, each other.

18.3.1 The Electrolyte–Antioxidant Paradox

Electrolytes such as sodium, potassium, magnesium, and calcium salts are essential for osmotic balance and hydration efficiency.
Yet their ionic nature can catalyze oxidation or destabilize colorants by promoting redox reactions.
For example:

- Trace iron in magnesium chloride may degrade ascorbic acid.
- Calcium ions can flocculate polyphenols or anthocyanins.
- High sodium or phosphate concentration can shift pH, altering pigment hue.

Therefore, electrolyte systems must be carefully balanced and buffered to maintain both functional efficacy and antioxidant protection.

18.3.2 Buffer Systems for Dual Stability

Buffers maintain pH within the optimal window for both antioxidants and colorants (usually pH 3.6–4.2).

Table 18.3.2: Dual-Function Buffer Systems for Antioxidant–Electrolyte Beverage Stability

Buffer System	Optimal pH Range	Function	Notes
Citric / Sodium Citrate	3.5–4.0	Maintains color stability, prevents oxidation	Ideal for anthocyanin-rich beverages
Malic / Sodium Malate	3.2–4.0	Provides soft acidity, taste brightness	Enhances fruit flavor; synergistic with ascorbate
Phosphate (Monobasic / Dibasic)	4.5–6.0	Controls buffering for neutral systems	Avoid in colored, polyphenol-rich drinks
Lactic / Sodium Lactate	3.5–4.2	Natural-label acidulant	Adds mild dairy note; suitable for clear hydration
Gluconic / Sodium Gluconate	3.8–5.0	Chelates metals, stabilizes antioxidants	Excellent for mineral waters

Formulator's Insight:

Use mixed buffers (citric + gluconate) for dual stability, citric controls pH, gluconate binds trace metals.

18.3.3 Chelation and Metal Ion Management

Electrolyte salts often carry trace metals that catalyze oxidation.
Chelating agents bind these ions, preventing degradation of antioxidants and pigments.

Table 18.3.3: Metal Chelation Strategies for Electrolyte–Antioxidant Systems

Chelator	Source / Label Status	Recommended Level	Function

Citric Acid	Natural fruit acid	0.1–0.2%	Mild chelation + pH control
Sodium Gluconate	Fermentation-derived	0.1–0.3%	Strong metal chelator; enhances clarity
EDTA (Calcium Disodium)	Synthetic	10–30 ppm	Potent stabilizer (if permitted)
Phytic Acid	Rice / soy derived	0.02–0.05%	Natural alternative; clean label compatible

Best Practice:

Always verify that total calcium + magnesium concentrations do not exceed 200 ppm when anthocyanins are used, higher levels risk precipitation or color dulling.

18.3.4 Antioxidant Network Design

A multi-layered antioxidant system provides superior protection than a single compound because degradation mechanisms are sequential, primary radicals, peroxides, and metal-catalyzed oxidation each require different interceptors.

Table 18.3.4: Antioxidant Network Design for Functional Hydration Beverages

Antioxidant Type	Phase	Example	Function
Primary (radical scavengers)	Water	Ascorbic acid, polyphenols	Neutralize free radicals
Secondary (regenerative)	Oil	Tocopherols, rosemary extract	Regenerate primary antioxidants
Chelating (preventive)	Water	Citric acid, gluconate	Bind metals that trigger oxidation
Synergists	Both	Lecithin, ascorbyl palmitate	Bridge lipid and aqueous phases

Optimal Ratio (Guideline):

- Ascorbic acid: Tocopherol : Citric acid = 1 : 0.5 : 0.2 (by weight basis relative to total solids).

This balanced system slows both aqueous and lipid oxidation, extending color and flavor stability.

18.3.5 Managing Vitamin C in Electrolyte Systems

Ascorbic acid is an excellent antioxidant but highly reactive in the presence of metal ions.

To preserve potency:

- Maintain pH 3.5–3.8.
- Add ascorbic acid post-dissolution of salts to minimize early oxidation.
- Use nitrogen-flushed blending tanks.
- Pair with sodium citrate or gluconate to complex metals.
- For extended shelf life, consider ascorbyl phosphate (stable derivative).

Example:

A lemon–mint electrolyte drink containing 200 ppm ascorbic acid and 0.25% sodium citrate retained 92% vitamin C activity after 9 months in amber PET (ambient).

18.3.6 Synergy Between Polyphenols and Electrolytes

Polyphenols such as catechins or flavonoids enhance antioxidant power, but excessive ionic strength can cause haze or precipitation.
To avoid this:

- Keep total dissolved solids ≤3.5%.
- Maintain polyphenol content <0.15% for clear drinks.
- Add polyphenols after cooling (below 40 °C).
- Incorporate low-acyl gellan gum (0.02–0.05%) for colloidal stability.
- Use ultrafiltration (0.2 μm) to remove insoluble complexes.

Sensory Enhancement:

Low electrolyte–polyphenol systems have a crisp, mineral-fruit flavor; higher ones taste metallic, balance acidity and sweetness to neutralize metallic hints.

18.3.7 Example Formulation - Antioxidant-Enhanced Electrolyte Beverage

Table 18.3.7: Example Formulation of Antioxidant-Enhanced Electrolyte Beverage

Ingredient	Function	% w/w
Purified water	Solvent	q.s. to 100
Glucose	Energy / osmotic balance	2.0
Sodium citrate	Buffer / electrolyte	0.25
Potassium chloride	Electrolyte	0.10
Magnesium gluconate	Electrolyte + chelator	0.05
Citric acid	pH 3.8 control	0.10
Ascorbic acid	Primary antioxidant	0.03
Tocopherol (Vit E emulsion)	Secondary antioxidant	0.02
Polyphenol extract (green tea)	Functional antioxidant	0.05
Stevia (Reb M)	Sweetness	0.02
Natural flavor (citrus-berry)	Sensory	0.03
Gum Arabic	Stabilizer	0.05

Performance Metrics:

- pH 3.8 ± 0.05
- Osmolality ≈ 285 mOsm/kg
- Clarity < 1 NTU

- Color retention > 90% (6 months)
- Vitamin C retention > 85%
- Shelf life ≥ 9 months (ambient, UV-protected)

18.3.8 Processing and Packaging Recommendations

1. Dissolve electrolytes and buffers first to ensure ionic balance.
2. Add antioxidants and polyphenols last, under gentle mixing (<45 °C).
3. Maintain oxygen <1 ppm during filling (nitrogen headspace).
4. Use amber PET or aluminum cans for light-sensitive formulas.
5. If hot-filled, limit temperature to ≤ 85 °C / < 30 s.
6. Store finished product at 15–25 °C.

18.3.9 Formulation Insight

- Treat electrolyte balance and antioxidant preservation as a single integrated system.
- Chelation and pH control are the twin anchors of stability.
- Avoid excessive mineral hardness, clarity and taste suffer before the body benefits.
- Proper antioxidant layering enables a clean, bright, and shelf-stable beverage that maintains both functional efficacy and visual appeal.

19. Plant Extracts, Polyphenols, and Functional Botanicals

Plant-derived actives are the cornerstone of functional hydration and recovery beverages. They provide a broad spectrum of benefits, from antioxidant protection and immune support to mood modulation and endurance enhancement. However, their natural complexity introduces formulation challenges: variable potency, taste bitterness, instability under acidic conditions, and compatibility issues with electrolytes or proteins.

This chapter provides a practical framework for selecting, standardizing, and stabilizing botanical and polyphenolic actives for clear, appealing, and effective functional beverage systems.

19.1 Bioactive Plant Compounds in Hydration and Recovery Systems

Bioactive plant compounds add a powerful layer of functionality to hydration and recovery systems, contributing antioxidant protection, metabolic support, anti-inflammatory effects, and cognitive or physiological benefits depending on their origin and chemistry. These botanicals, ranging from polyphenols and adaptogens to herbal extracts and concentrated phytonutrients, interact dynamically with acids, electrolytes, and flavor systems, influencing both stability and sensory expression. Understanding their roles and integration requirements allows formulators to design beverages that deliver targeted physiological advantages while maintaining clarity, balance, and shelf-life reliability.

19.1.1 Functional Role of Botanical Extracts

Plant extracts add biochemical functionality and natural appeal to hydration beverages.
Their actions often include:

- Antioxidant protection: scavenging free radicals and supporting recovery.

- Vascular enhancement: improving microcirculation and oxygen delivery.
- Anti-inflammatory modulation: reducing exercise-induced muscle stress.
- Adaptogenic balance: supporting stress resilience and focus.
- Immunomodulation: strengthening cellular defense mechanisms.

Key Performance Advantage:

Unlike synthetic actives, botanicals often act synergistically, multiple compounds working in harmony to modulate physiological responses while also reinforcing sensory depth.

19.1.2 Major Classes of Functional Botanicals

Table 19.1.2: Functional Botanical Classification Matrix

Botanical Group	Principal Compounds	Core Function	Typical Inclusion Range	Formulation Considerations
Polyphenol-rich Extracts	Catechins, flavonoids, anthocyanins	Antioxidant, recovery, anti-fatigue	0.02–0.10%	Sensitive to heat, light, and pH >4
Adaptogenic Herbs	Withanolides, ginsenosides, rosavins	Stress balance, endurance, mood	0.02–0.05%	Bitterness; mask with citrus, spice
Immune Botanicals	Echinacea, elderberry, green tea	Immune modulation, antioxidant	0.05–0.15%	Maintain pH 3.5–4.0
Digestive / Detox Herbs	Ginger, mint, turmeric	Anti-inflammatory, circulation	0.02–0.10%	Volatile oils; encapsulate for stability

Cognitive Botanicals	Bacopa, guarana, gotu kola	Focus, cognition	0.01–0.03%	Use standardized extracts for precision
Flavor Botanicals	Lemongrass, hibiscus, rooibos	Natural flavor and function	0.05–0.20%	Naturally colorant-active

19.1.3 Polyphenols and Their Hydration Benefits

Polyphenols are powerful antioxidants found in tea, berries, cocoa, and grape extracts.
In hydration systems, they:

- Reduce oxidative muscle stress during exercise.
- Enhance nitric oxide bioavailability for improved blood flow.
- Stabilize vitamin C and natural pigments via redox cycling.
- Contribute mild astringency, which enhances freshness perception.

Key Polyphenolic Families and Their Functions:

Table 19.1.3: Key Polyphenolic Families in Hydration Beverages - Sources, Functions, and Sensory Profiles

Polyphenol Type	Source	Function	pH Stability	Flavor Profile
Catechins (EGCG)	Green tea	Antioxidant, metabolism	3.5–4.0	Bitter, vegetal
Anthocyanins	Berries, purple corn	Color + antioxidant	3.0–4.0	Fruity, tart
Proanthocyanidins (OPCs)	Grape seed, cocoa	Vascular protection	3.5–5.0	Dry, tannic

Curcuminoids	Turmeric	Anti-inflammatory	3.5–6.0	Warm, earthy
Flavonols (Quercetin, Rutin)	Citrus peel, onion	Anti-inflammatory, antioxidant	3.5–4.5	Bitter, aromatic

Formulation Tip:

Balance bitterness from polyphenols with mild sweetness (isomaltulose, erythritol) and bright acidity (citric + malic).
Natural masking agents such as vanilla or stevia (Reb M) smooth the astringency without added calories.

19.1.4 Botanical Extract Standardization and Labeling

Consistency and regulatory compliance depend on extract standardization, ensuring defined levels of bioactive markers.

Table 19.1.4: Standardization Specifications for Beverage-Grade Botanical Extracts

Botanical	Standard Marker	Target Concentration	Typical Dosage in Beverages
Green Tea Extract	EGCG	40–50%	100–150 mg/serving
Grape Seed Extract	OPCs	90–95%	50–100 mg/serving
Ashwagandha	Withanolides	5%	150–300 mg/serving
Rhodiola rosea	Rosavin + salidroside	3% + 1%	100–200 mg/serving
Turmeric	Curcuminoids	95%	50–100 mg/serving
Elderberry	Anthocyanins	15%	100–150 mg/serving

Bacopa monnieri	Bacosides	20%	100 mg/serving

Labeling Guidance:

- Declare standardized actives on the ingredient list.
- Use consumer-friendly descriptors ("with standardized ashwagandha root extract") rather than chemical names.
- Comply with regional labeling requirements (EU, FDA, GCC) regarding botanical extracts in beverages.

19.1.5 Stability and Compatibility with Beverage Matrices

Botanical actives interact strongly with beverage components. The matrix, water, acid, salts, and flavor system, determines stability and clarity.

Key Considerations:

- pH: Most botanicals stable at 3.6–4.0.
- Ionic strength: High electrolytes may cause clouding, use stabilizers like gum arabic (0.05%).
- Thermal sensitivity: Avoid >85 °C exposure.
- Oxygen management: Nitrogen flush recommended during blending.
- Light sensitivity: Amber or UV-block PET essential for polyphenol retention.
- Flavor volatility: For oils (mint, ginger), use encapsulated or emulsified extracts.

19.1.6 Example Formulation - Botanical Polyphenol Hydration Beverage

Table 19.1.6: Botanical Polyphenol Hydration Beverage Formula

Ingredient	Function	% w/w
Purified water	Solvent	q.s. to 100

Sodium citrate	Buffer + electrolyte	0.25
Potassium chloride	Electrolyte	0.10
Citric acid	pH 3.8 control	0.10
Green tea extract (50% EGCG)	Polyphenol antioxidant	0.03
Grape seed extract (95% OPCs)	Vascular support	0.02
Isomaltulose	Mild sweetness + energy	2.0
Ascorbic acid	Vitamin co-antioxidant	0.03
Natural flavor (lemon–mint)	Sensory	0.03
Stevia (Reb M)	Sweetener	0.02
Gum Arabic	Stabilizer	0.05

Performance Metrics:

- pH: 3.8
- Clarity: <1 NTU
- Polyphenol retention: >85% after 6 months
- Color: Pale green-gold (clear)
- Shelf life: 9–12 months
- Flavor: Light, refreshing, mild green-tea bitterness balanced by citrus brightness

Formulation Insight

- Botanical and polyphenol systems must be built around their chemistry, not added as decorative actives.
- Stability is driven by pH, metal control, and oxygen exclusion, protect these parameters at every production stage.
- Use sensory pairing (acid + aroma + natural sweetness) to turn bitterness into perceived freshness.

- The clean-label opportunity lies in combining plant identity, physiological function, and sensory harmony within one beverage architecture.

19.2 Extraction Methods, Standardization, and Quality Control for Functional Botanicals

The value of a botanical ingredient depends not only on its origin but also on how it is extracted, standardized, and verified.
Extraction methods determine the bioactive profile, flavor intensity, and stability of plant-derived materials in beverages.
To achieve batch-to-batch consistency and regulatory acceptance, beverage developers must understand how extraction conditions, solvents, and drying processes influence functional and sensory performance.

This section explains the key extraction technologies, standardization principles, and quality control (QC) parameters essential for beverage-grade botanical actives.

19.2.1 Principles of Botanical Extraction

Extraction isolates bioactive compounds (polyphenols, alkaloids, terpenes, flavonoids) from plant matrices using solvents, temperature, and mechanical techniques.
In beverage applications, the ideal extract should be:

- Water-soluble
- Standardized in active content
- Low in color variability
- Sensory balanced (non-bitter or astringent)
- Microbiologically safe

Extraction Goals:

1. Concentrate desirable actives.
2. Eliminate undesirable components (fats, waxes, tannins).
3. Maintain chemical integrity (no oxidation or hydrolysis).

4. Produce a stable, dispersible powder or liquid for formulation.

19.2.2 Common Extraction Methods for Beverage-Grade Botanicals

Table 19.2.2: Botanical Extract Standardization Parameters - Solubility, Bioactives, and Stability Metrics

Extraction Method	Solvent System	Typical Applications	Advantages	Limitations
Aqueous Extraction	Water	Green tea, hibiscus, ginger	Clean-label, solvent-free	Limited solubility for non-polar actives
Hydroalcoholic Extraction	Water + ethanol (20–80%)	Adaptogens, ginseng, rhodiola	Broad-spectrum extraction	Requires ethanol removal; regulatory limits
Supercritical CO_2 Extraction	CO_2 (at high pressure)	Turmeric, coffee, citrus oils	Solvent-free, selective for lipophilic compounds	High cost, equipment-intensive
Ultrasound-Assisted Extraction (UAE)	Water or ethanol	Polyphenols, flavonoids	Accelerated yield, energy-efficient	Scale-up complexity
Microwave-Assisted Extraction (MAE)	Water/ethanol mix	Polyphenols, anthocyanins	Short time, high efficiency	Risk of thermal degradation

Enzymatic Extraction	Cellulase, pectinase + water	Fruit peels, adaptogens	Gentle release, higher clarity	Enzyme cost, long time
Spray-Dried Extracts	(drying step)	All botanical powders	Excellent shelf life and solubility	Slight loss of volatile components

Formulator's Insight:

For beverages, aqueous or hydroalcoholic extracts are most common, as they provide the best combination of solubility, clean label compliance, and flavor balance.

19.2.3 Standardization of Bioactive Components

Standardization ensures each batch of botanical extract delivers consistent physiological potency.
This process involves quantifying one or more marker compounds, either the main bioactive or a reliable proxy for extract quality.

Approaches to Standardization:

- Marker-based: defining a specific compound (e.g., 5% withanolides in ashwagandha).
- Total-activity-based: measuring total polyphenol or flavonoid content (e.g., ≥40% catechins).
- Fingerprint-based: chromatographic pattern matching via HPLC or LC-MS to maintain phytochemical balance.

Table 19.2.3: Standardization Specifications for Beverage-Grade Botanical Extracts

Botanical	Marker Compound	Typical Standardization	Assay Method

Green Tea Extract	EGCG	40–50%	HPLC, UV spectrophotometry
Grape Seed Extract	OPCs	90–95%	Folin–Ciocalteu, HPLC
Ashwagandha	Withanolides	5–7%	HPLC
Rhodiola rosea	Rosavin + salidroside	3% + 1%	HPLC
Turmeric	Curcuminoids	90–95%	UV/Vis
Elderberry	Anthocyanins	15–20%	pH differential method
Ginseng	Ginsenosides	10–20%	TLC / HPLC

Regulatory Note:

When using standardized botanical extracts, declare the active marker and percentage on product specifications for international compliance (EU, FDA, EFSA, GCC, ASEAN).

19.2.4 Solubility and Form Adaptation for Beverage Applications

Botanical extracts are produced in several physical forms, each with distinct formulation implications.

Table 19.2.4: Physical Forms of Botanical Extracts for Beverage Applications

Form	Description	Solubility	Best Use
Spray-Dried Powder	Dried aqueous extract with maltodextrin	Excellent	Dry blending and reconstitution
Liquid Extract (10–20% solids)	Concentrated decoction	Good	Direct liquid dosing

Hydroalcoholic Extract (Soft)	Semi-viscous	Moderate	Functional tonics, not clear beverages
Emulsified Extract	Lipid-soluble actives in aqueous carriers	Excellent	Turmeric, ginger, citrus oils
Microencapsulated Extract	Encased in carbohydrates or proteins	Excellent	Shelf-stable clear functional drinks

Formulation Tip:

Select extract forms based on target beverage type:

- Clear isotonic drinks: use microencapsulated or spray-dried extracts.
- Cloudy or smoothie-style drinks: use liquid or emulsified extracts.
- Carbonated drinks: avoid unencapsulated polyphenol extracts (foam instability risk).

5. Quality Control (QC) and Analytical Verification

Consistent quality requires rigorous testing of both raw extracts and finished beverages.

19.2.5 QC Parameters and Recommended Methods:

Table 19.2.5: Quality Control Parameters for Botanical Extracts

Parameter	Purpose	Analytical Method	Target
Identity Verification	Confirm botanical species	HPTLC or DNA barcoding	Match reference profile
Active Marker Assay	Quantify potency	HPLC or UV/Vis	Within ±10% label claim
Moisture Content	Prevent microbial growth	Karl Fischer or LOD	<5% (powders)

Microbial Load	Ensure safety	Total plate count	$<10^3$ CFU/g
Heavy Metals	Ensure purity	ICP-MS	Pb <3 ppm, As <1 ppm
Pesticide Residues	Regulatory compliance	GC–MS	Must meet regional limits
Solvent Residues	For hydroalcoholic extracts	GC	Ethanol <0.5%
Color / Clarity / pH	Product uniformity	Visual + pH meter	Consistent per batch

19.2.6 Ensuring Batch-to-Batch Consistency

Consistency depends on traceable sourcing, controlled extraction, and robust analytical validation.

Best Practices for Beverage Manufacturers:

1. Qualified Supplier Program – Audit extraction process and solvent system.
2. Certificate of Analysis (CoA) – Must list active percentage, heavy metal results, and microbial data.
3. Retention Samples – Keep at least 6 months for verification.
4. Standardization Reference Library – Maintain internal HPLC fingerprints.
5. Sensory QC – Evaluate aroma and flavor intensity at standardized dilution.

Formulator's Insight:

A consistent extract guarantees regulatory security, predictable sensory impact, and reliable biological efficacy, the three pillars of functional beverage credibility.

19.2.7 Example Process Flow – Botanical Extract Integration

1. Raw Material Verification: Authenticate plant source and active content.
2. Extraction: Perform aqueous or hydroalcoholic extraction under controlled pH (5–7).

3. Filtration: Clarify and remove insoluble residues (0.45 μm).
4. Concentration: Vacuum-evaporate to 10–20% solids.
5. Drying: Spray-dry or encapsulate to enhance stability.
6. Blending: Mix into beverage base post-buffer addition, below 45°C.
7. Packaging: Use UV-protective containers with nitrogen flush.
8. Final Testing: Validate pH, clarity, color, and potency retention.

19.2.8 Example Specification – Beverage-Grade Ashwagandha Extract

Table 19.2.8: Example Specification of Beverage-Grade Ashwagandha Extract

Parameter	**Specification**
Appearance	Light tan powder
Solubility	100% water soluble
Active Marker	5% withanolides (HPLC verified)
Moisture	<5%
pH (1% solution)	4.0–6.0
Odor	Mild herbal
Microbial Load	<1000 CFU/g
Heavy Metals	Within regulatory limits
Packaging	Foil-sealed, nitrogen flushed
Shelf Life	24 months (cool, dry storage)

Formulation Insight

- Extraction defines bioefficacy and taste, a poor extract cannot be fixed by formulation.

- Always select suppliers with standardized bioactives and transparent QC data.
- Match extract chemistry to your beverage process (pH, temperature, ionic strength).
- For long-term stability, integrate encapsulation and chelation systems to protect sensitive botanicals from degradation.
- Authentic plant actives, when properly extracted and standardized, differentiate premium hydration products from commodity formulations.

19.3 Polyphenol Stability, Solubility, and Sensory Management in Beverage Systems

Polyphenols are among the most sought-after actives in functional beverages due to their antioxidant, anti-inflammatory, and vascular-support properties.
However, they are also among the most unstable, susceptible to oxidation, precipitation, and flavor drift in aqueous systems.
Maintaining clarity, taste, and bioactivity requires understanding their chemistry and controlling the beverage environment that surrounds them.

19.3.1 Chemical Nature of Polyphenols

Polyphenols are a diverse class of plant secondary metabolites characterized by multiple hydroxyl groups attached to aromatic rings.
These structures make them powerful antioxidants but also highly reactive, easily oxidized in the presence of light, heat, oxygen, or trace metals.

Major Polyphenol Families in Beverage Formulation:

Table 19.3.1: Major Polyphenol Families in Beverage Formulation - Chemistry, Function, and Flavor Impact

Family	Typical Source	Key Compounds	Functional Role	Sensory Traits

Flavanols	Green tea, cocoa	Catechin, EGCG	Antioxidant, metabolism	Bitter, astringent
Flavonols	Citrus, onion, buckwheat	Quercetin, rutin	Anti-inflammatory	Bitter, herbal
Anthocyanins	Berries, purple corn	Cyanidin, malvidin	Antioxidant + color	Tart, fruity
Phenolic Acids	Coffee, fruit	Caffeic, chlorogenic acid	Antioxidant, metabolism	Mild bitterness
Proanthocyanidins (OPCs)	Grape seed, pine bark	Oligomeric catechins	Vascular health	Dry, tannic
Curcuminoids	Turmeric	Curcumin, demethoxycurcumin	Anti-inflammatory	Earthy, warm

19.3.2 Polyphenol Solubility and pH Behavior

Polyphenol solubility depends on hydroxyl density, polymer size, and pH.

Table 19.3.2: Polyphenol Solubility and pH Behavior - Stability Windows, Ionic Constraints, and Processing Guidelines for Functional Beverages

Solubility Factor	Impact	Practical Guidance
pH	Polyphenols are most stable and soluble at pH 3.5–4.2	Maintain beverage pH within this range
Ionic Strength	High electrolyte levels reduce solubility (ionic shielding)	Limit total salts <0.5%

Temperature	Heat accelerates oxidation	Keep processing $\leq 85°C$
Oxygen	Dissolved O_2 triggers browning	Use nitrogen sparging and tight headspace control
Complexation	Polyphenols bind proteins and metals	Use chelators and microfiltration (0.2 μm)

Formulation Tip:

Add polyphenols after cooling and pH adjustment to minimize degradation during thermal or acidic exposure.

19.3.3 Mechanisms of Polyphenol Degradation

a. Oxidation:

Polyphenols auto-oxidize in the presence of oxygen and metal ions, leading to color browning and loss of antioxidant capacity.
Solution: Pair with ascorbic acid, citric acid, and tocopherols to regenerate antioxidant activity.

b. Polymerization:

Oxidized polyphenols can polymerize, causing haze or sediment.
Solution: Maintain low dissolved oxygen (<1 ppm) and avoid high divalent cation concentration (Ca^{2+}, Mg^{2+}).

c. Hydrolysis:

Under high pH (>5), phenolic esters can hydrolyze, reducing stability.
Solution: Maintain acidified pH 3.5–4.0 and use buffered systems.

19.3.4 Stabilization Strategies

Table 19.3.4: Polyphenol Stabilization Matrix - pH Control, Chelation, Antioxidant Synergy, and Encapsulation for Long-Term Clarity and Potency

Strategy	Mechanism	Key Ingredients / Techniques

pH Control	Keeps polyphenols in stable, non-ionized form	Citric or malic acid buffers (pH 3.8–4.0)
Chelation	Prevents metal-catalyzed oxidation	Citric acid, sodium gluconate, phytic acid
Antioxidant Synergy	Regenerates oxidized polyphenols	Ascorbic acid + tocopherol system
Encapsulation	Physically isolates actives from oxygen	Maltodextrin, gum arabic, β-cyclodextrin
Oxygen Exclusion	Limits oxidative reactions	Nitrogen or CO_2 flushing during blending
Thermal Control	Reduces degradation during filling	Cold-fill or short hot-fill ≤85°C
UV Protection	Prevents photodegradation	Amber PET, aluminum, or opaque sleeve packaging

Formulator's Insight:

Anthocyanin-rich or polyphenol-heavy systems benefit greatly from oxygen scavengers in packaging (e.g., ascorbate-lined caps).

19.3.5 Sensory Management: Reducing Bitterness and Astringency

Polyphenols naturally interact with salivary proteins, causing the dry, puckering mouthfeel known as astringency.
While beneficial in teas and wines, excessive astringency is undesirable in hydration beverages.

Sensory Balancing Techniques:

Table 10.3.5: Sensory Management Strategies for Reducing Bitterness and Astringency in Polyphenol Systems

Challenge	Solution	Example
Bitterness	Use natural sweetness (erythritol, isomaltulose) or masking agents (vanillin, glycine)	Lemon–green tea blend
Astringency	Add low-viscosity hydrocolloids (pectin 0.05–0.1%) or plant glycerol	Grape seed beverage
Metallic notes	Use citric acid + sodium citrate buffer	Caffeine–polyphenol combinations
Flat flavor	Use acid layering (citric + malic) to add brightness	Hibiscus–berry formulations
Flavor dulling during storage	Incorporate light antioxidants and natural essential oil notes	Mint or citrus top-notes

Taste Architecture Rule:

Balance three layers, acid brightness, natural sweetness, and functional bitterness, to create a sophisticated but drinkable flavor.

19.3.6 Polyphenol–Electrolyte Compatibility

Polyphenols can react with divalent ions such as Ca^{2+} and Mg^{2+} to form insoluble complexes.
To ensure clarity:

- Limit Ca^{2+} and Mg^{2+} below 200 ppm.
- Prefer sodium citrate over calcium phosphate buffers.
- Add polyphenols post-salt dissolution.
- Use low-acyl gellan gum (0.02–0.05%) for colloidal protection.

Example:

In a green tea–electrolyte drink, using sodium citrate (0.25%) instead of calcium chloride improved clarity (NTU <1.0) and color stability (+90%) after 6 months.

19.3.7 Example Formulation - Polyphenol-Rich Hydration Beverage

Table 19.3.7: Polyphenol-Rich Hydration Beverage - Dual-Extract Antioxidant System (Green Tea + Grape Seed), pH 3.8, NTU <1.0

Ingredient	Function	% w/w
Purified water	Solvent	q.s. to 100
Citric acid	pH 3.8 control	0.10
Sodium citrate	Buffer / electrolyte	0.25
Potassium chloride	Electrolyte	0.10
Green tea extract (40% catechins)	Antioxidant / active	0.03
Grape seed extract (95% OPCs)	Vascular support	0.02
Ascorbic acid	Antioxidant stabilizer	0.03
Natural flavor (lemon–mint)	Sensory	0.03
Erythritol	Natural sweetness	0.40
Gum Arabic	Stabilizer / mouthfeel	0.05

Performance Metrics:

- pH: 3.8
- Clarity: <1.0 NTU
- Color retention: >90% (6 months)
- Polyphenol activity: >85% retained
- Taste: Crisp, refreshing, mildly astringent citrus-green tea note

19.3.8 Packaging and Shelf-Life Optimization

- Light protection: Use amber or UV-blocked PET.
- Headspace management: Nitrogen flush to remove oxygen (<1 ppm).
- Cap selection: Use low-permeability closures with oxygen barrier liners.
- Storage conditions: 15–25 °C, avoid direct sunlight.
- Shelf life: Typically, 9–12 months ambient or 18 months under aseptic cold-fill.

19.3.9 Formulation Insight

- Polyphenols are dynamic systems, they need stabilization chemistry, not just formulation.
- Always integrate acid–antioxidant–chelator triads for clarity and potency.
- Taste control defines success; chemical stability alone doesn't ensure consumer acceptance.
- The best polyphenol beverages are balanced, bright, and functional, delivering sensory pleasure alongside measurable wellness benefits.

19.4 Integration of Botanical Extracts with Flavor and Aroma Systems

Botanical extracts bring complexity, authenticity, and functionality to modern hydration beverages, yet their strong flavors, volatiles, and color-active components must be carefully balanced within the flavor matrix.
A successful integration achieves functional authenticity without compromising drinkability and shelf stability.

This section explores how to design flavor–botanical systems that express natural identity, maintain clarity, and deliver multi-sensory coherence.

19.4.1 The Flavor–Botanical Relationship

Every botanical extract carries its own sensory fingerprint, often herbal, earthy, bitter, or aromatic.

In beverage systems, these profiles interact with acids, sweeteners, and flavor carriers to create the final sensory impression.

Table 19.4.1: Botanical Extract Sensory Profiles - Flavor Archetypes, Functional Positioning, and Optimal Flavor Pairings

Botanical Type	Sensory Traits	Functional Theme	Best Flavor Synergies
Green Tea / Matcha	Bitter, grassy, umami	Focus, metabolism	Citrus, mint, pear, jasmine
Ginseng / Ashwagandha	Earthy, slightly bitter	Endurance, stress balance	Vanilla, cocoa, spice, caramel
Rhodiola / Bacopa	Woody, resinous	Cognitive support	Berry, hibiscus, tropical
Turmeric / Ginger	Warm, spicy, aromatic	Anti-inflammatory, energy	Lemon, honey, mango, mint
Elderberry / Hibiscus	Tart, fruity	Immune, antioxidant	Blackcurrant, rose, pomegranate
Mint / Tulsi (Holy Basil)	Cooling, refreshing	Relaxation, balance	Lime, cucumber, lemongrass

Formulator's Insight:

Match the emotional and functional narrative of each botanical with complementary flavor notes, for example, "Calm Focus" → Green tea + mint + ashwagandha or "Daily Defense" → Elderberry + citrus + turmeric.

19.4.2 Sensory Mapping and Flavor Layering

Successful botanical beverages rely on layered flavor architecture, designed to manage intensity, masking, and aromatic lift.

Three-Layer Flavor Model:

Table 19.4.2: Flavor Layering Architecture for Botanical Beverages - Top, Middle, and Base Notes with Intensity Management Guidelines

Layer	Function	Example Components
Top Notes	First aromatic impact (volatile compounds)	Citrus zest, mint, ginger, lemongrass
Heart Notes	Mid-palate character (main flavor identity)	Berry, tea, herbal, floral extracts
Base Notes	Depth and aftertaste balance	Vanilla, caramel, cocoa, woody tones

Balancing Strategy:

- Use bright acid–top note pairs (citric + lemon oil) to mask bitterness.
- Integrate rounding base notes (vanilla, glycerol, stevia) to smooth astringency.
- Avoid over-layering, keep total flavor oil load ≤0.2% for clear beverages.

Example:

A "Ginger–Turmeric Recovery" drink layers lemon top notes for freshness, ginger and turmeric for warmth, and honey–vanilla base notes for smoothness.

19.4.3 Aroma Retention and Volatile Stability

Botanical volatiles (e.g., menthol, citral, eugenol) can easily evaporate or oxidize, leading to flavor loss or aroma drift.
Managing this requires controlled emulsification, microencapsulation, and packaging protection.

Table 19.4.3: Aroma Stabilization Technologies for Botanical Volatiles - Emulsification, Encapsulation, and Packaging Strategies

Method	Function	Recommended Use
Emulsification (oil-in-water)	Stabilizes essential oils	0.05–0.2% natural flavor oils

Gum Arabic or Modified Starch	Encapsulates aroma compounds	3:1 carrier-to-oil ratio
Cold-Fill or Aseptic Process	Preserves volatiles	Use <45 °C blending
Light-Barrier Packaging	Protects aroma compounds from oxidation	Amber PET or shrink sleeve
Antioxidant Coupling	Stabilizes terpenes and aldehydes	Add ascorbic acid 0.02–0.05%

Formulation Tip:

Essential oils oxidize faster in alkaline or metal-rich environments; buffer pH around 3.8 and avoid high calcium salts.

19.4.4 Managing Flavor–Botanical Interactions

Certain flavor compounds can enhance or suppress the perception of botanical bitterness or astringency.

Table 19.4.4: Flavor–Botanical Interaction Matrix - Bitterness, Astringency, and Earthiness Masking Strategies with Functional Pairings

Botanical Taste Challenge	Flavor Strategy	Example Pair
Bitterness (e.g., ashwagandha)	Use natural vanillin, cocoa, or cinnamon for warmth	Vanilla–spice chai hydration
Astringency (polyphenols)	Round with malic acid and erythritol	Green tea–citrus blend
Earthiness (adaptogens)	Add bright fruit acids or herbal top notes	Ginseng–grapefruit–mint
Herbal Sharpness	Blend with floral middle notes	Tulsi–jasmine–lime

Spice Volatility	Anchor with mild sweetness (honey solids or isomaltulose)	Ginger–honey–lemonade

Formulator's Insight:

Low-level bitterness can enhance perceived "authenticity." Don't eliminate it entirely, instead, contextualize it through complementary aroma and sweetness.

19.4.5 Solubility and Emulsion Management

Many botanical flavors contain lipophilic components that require emulsifiers for stable dispersion in water.

Emulsion Design Guidelines:

- Oil load: 0.05–0.15%
- Emulsifier: Gum arabic (0.3–0.5%) or modified starch (0.2–0.4%)
- Homogenization: 50–70 MPa, 2 passes for fine droplet size (<0.3 μm)
- Antioxidant: Tocopherol or ascorbic acid to protect terpenes
- pH range: 3.5–4.0 for long-term clarity

Clarity Optimization:

For clear beverages, pre-filter emulsions through 0.45 μm before blending.

19.4.6 Example Formulation - Botanical Citrus–Mint Hydration Beverage

Table 19.4.6: Botanical Citrus–Mint Hydration Beverage Formulation

Ingredient	**Function**	**% w/w**
Purified water	Solvent	q.s. to 100
Citric acid	Acidity control	0.10
Sodium citrate	Buffer + electrolyte	0.25

Potassium chloride	Electrolyte	0.10
Green tea extract (40% catechins)	Antioxidant, metabolism	0.03
Ashwagandha extract (5% withanolides)	Stress modulation	0.02
Natural flavor oil (lemon–mint)	Top note	0.05
Gum Arabic	Emulsifier / stabilizer	0.05
Erythritol	Sweetness + mouthfeel	0.40
Ascorbic acid	Antioxidant	0.03
Stevia (Reb M)	Natural sweetener	0.02

Performance Metrics:

- pH: 3.8
- Clarity: <1 NTU
- Aroma retention: >85% after 6 months
- Taste: Refreshing, crisp, balanced botanical tone
- Shelf life: 9–12 months (cold-fill, light-protected)

19.4.7 Packaging and Aroma Retention Strategies

1. Use multilayer PET or aluminum bottles for volatile protection.
2. Minimize headspace to reduce oxygen exposure (<1% volume).
3. Apply shrink sleeves or UV-block coatings for visual and light shielding.
4. Include antioxidant pairing (ascorbic acid + tocopherol) for aroma preservation.
5. Store below 30 °C; avoid thermal cycling during distribution.

Example:

A citrus–mint hydration drink retained >90% limonene and menthol volatiles for 9 months when filled in nitrogen-flushed amber PET.

19.4.8 Formulation Insight

- The best functional beverages achieve sensory integrity; every ingredient serves both taste and purpose.
- Integrate botanicals as functional flavor anchors rather than hidden actives.
- Control pH, emulsification, and antioxidant support to ensure aromatic longevity.
- Always design with multisensory intent, color, aroma, flavor, and function should tell the same story.

19.5 Thermal, pH, and Storage Stability of Botanical Systems in Functional Drinks

Botanical compounds, especially polyphenols, flavonoids, and essential oils, are highly reactive molecules.
During beverage processing, they undergo oxidation, hydrolysis, or polymerization, resulting in color fading, bitterness, or loss of efficacy.

Understanding their thermal, pH, and storage behavior allows formulators to design process conditions and stabilization systems that ensure both functional integrity and sensory consistency.

19.5.1 Thermal Stability of Botanical Actives

Heat treatment is often required for beverage safety (pasteurization, hot-fill, or UHT).
However, most botanical actives are thermolabile, meaning they degrade when exposed to sustained high temperatures.

Typical Stability Ranges:

Table 19.5.1: Thermal Degradation Profiles of Botanical Actives

Botanical / Active	Thermal Stability	Optimal Process Condition	Degradation Pathway
Green Tea Catechins (EGCG)	Moderate	≤85 °C for ≤30 s	Oxidation / polymerization
Curcumin (Turmeric)	High (in emulsions)	≤90 °C	Isomerization to inactive forms
Ginsenosides (Ginseng)	Moderate	≤80 °C	Hydrolysis under acid
Withanolides (Ashwagandha)	Low–moderate	≤75 °C	Thermal oxidation
Anthocyanins (Berry pigments)	Low	≤75 °C	Degradation to chalcones
Volatile oils (Ginger, mint)	Low	≤60 °C	Evaporation / oxidation

Best Practices:

- Use short-time, low-temperature pasteurization (HTST) whenever possible.
- Add heat-sensitive actives after pasteurization in cold blending steps.
- For shelf-stable beverages, opt for aseptic cold-fill with nitrogen sparging.
- Include antioxidants (ascorbic acid, tocopherol) to slow thermal oxidation.

Formulator's Insight:

Thermal stress can reduce functional efficacy by 10–40% even if visual clarity remains intact, always re-test potency post-processing.

19.5.2 pH Sensitivity and Functional Optimization

Botanical stability is pH-dependent.
Acidic systems (pH 3.5–4.0) are optimal for beverage preservation and microbial control but can stress certain phytochemicals.

General pH Behavior by Class:

Table 19.5.2: pH-Dependent Stability of Botanical Classes

Botanical Class	Ideal pH Range	Reaction Outside Range	Stabilization Approach
Polyphenols	3.5–4.2	Oxidation at >5	Buffer + chelator
Anthocyanins	3.0–3.8	Color loss at >4.2	Acid stabilization
Adaptogenic Alkaloids	4.0–6.0	Hydrolysis at <3.5	Use microencapsulated forms
Volatile Terpenes	3.5–4.5	Hydrolysis / dull aroma	pH buffer + antioxidants
Flavonoids (Quercetin)	3.5–4.0	Precipitation at <3	Solubilized or nanoformulated extract

Buffering System Recommendation:

Use citric + sodium citrate (0.25–0.3%) or malic acid (0.1%) for pH balance and sensory brightness without harshness.

19.5.3 Combined Effect of pH and Heat

Many degradation reactions accelerate when both low pH and heat are applied simultaneously, particularly hydrolysis and oxidative browning.

Example Behavior:

- EGCG (green tea): stable for 6 months at 25 °C, but only 4 weeks at 40 °C / pH 3.5.
- Anthocyanins: retain color at pH 3.5 / 20 °C, but fade 40% at 80 °C.

- Turmeric curcuminoids: stable at pH 4.0 even at 90 °C due to hydrophobic encapsulation.

Mitigation Strategies:

1. Optimize sequence of addition: dissolve salts and acids first, add botanicals last.
2. Apply encapsulation: protect actives from protonation and heat exposure.
3. Reduce dissolved oxygen: oxidation doubles for every 10 °C rise.
4. Use polyphenol stabilizers: e.g., gum arabic 0.05–0.1% or pectin 0.02–0.05%.

19.5.4 Storage Stability and Shelf-Life Control

Even after filling, botanical beverages continue to evolve chemically.
Light, temperature, oxygen permeability, and headspace all determine the effective shelf life.

Table 19.5.4: Botanical Beverage Shelf-Life Control

Storage Stress	Common Outcome	Prevention Strategy
Light Exposure	Color fading, off-flavor	Amber PET, opaque labels, aluminum cans
Oxygen Diffusion	Polyphenol oxidation	Nitrogen flush, oxygen scavenger liners
High Temperature (>35 °C)	Accelerated degradation	Controlled logistics, cool storage
Moisture Migration (in powders)	Caking, potency loss	Foil barrier pouches, silica gel
Long Shelf Life (>12 mo)	Gradual loss of actives	Slight over-fortification at filling

Example:

A turmeric–ginger beverage with 0.03% ascorbic acid and 0.1% sodium citrate retained >90% curcuminoids and flavor after 12 months at 25 °C.

19.5.5 Accelerated Shelf-Life Testing (ASLT)

To predict product stability within practical timelines, accelerated testing at elevated temperature and humidity is standard.

Testing Protocol Example:

- 40 °C, 75% RH for 3 months → equivalent to ~12 months ambient.
- Evaluate color (ΔE), pH drift, polyphenol content, flavor retention.
- Acceptable limits:
    - ΔE < 2 (no visible color change)
    - pH drift < 0.1
    - Active retention > 85%
    - Sensory score ≥ 4/5 (panel test)

Analytical Techniques:

- HPLC for catechins, curcuminoids, or withanolides.
- Spectrophotometric color analysis (CIE Lab*).
- GC–MS for volatile retention (menthol, limonene).
- Headspace oxygen monitoring (<1 ppm target).

19.5.6 Packaging Optimization for Botanical Stability

Table 19.5.6: Packaging Selection Matrix for Botanical Stability

Packaging Type	Protection Strength	Ideal Application
Amber PET (multi-layer)	Excellent light + oxygen barrier	Polyphenol / tea beverages

Aluminum Can	Superior oxygen protection	Long-shelf-life functional drinks
Glass Bottle (dark)	Inert, premium	Short-distribution channels
Aseptic Carton	Good barrier, eco option	Neutral-pH or plant-milk bases
Foil Sachet (powder)	Excellent moisture protection	Dry beverage mixes

Headspace Control:

Maintain <1% oxygen by nitrogen sparging or CO_2 blanketing.
Optionally include oxygen scavengers in caps or liners for long-term protection.

19.5.7 Example: Thermal–pH–Stability Matrix for Common Botanicals

Table 19.5.7: Botanical Ingredient Stability Matrix

Botanical Extract	Thermal Stability	pH Stability	Recommended Process
Green Tea	Moderate	3.5–4.2	HTST (85 °C / 30 s), pH 3.8
Ashwagandha	Low	4.0–6.0	Cold-fill, pH 4.5
Rhodiola	Moderate	3.8–5.0	Short hot-fill, pH 3.9
Turmeric (Emulsified)	High	3.5–7.0	Hot-fill, pH 3.8
Elderberry	Low	3.0–3.8	Cold-fill, pH 3.6
Hibiscus	Moderate	3.0–3.8	HTST, pH 3.6
Ginger	Low	3.5–4.0	Cold-fill or low-temp pasteurization

19.5.8 Example Formulation - Heat- and Acid-Stable Botanical Beverage

Table 19.5.8: Heat- and Acid-Stable Botanical Beverage Formulation

Ingredient	Function	% w/w
Purified water	Solvent	q.s. to 100
Citric acid	pH 3.8 adjustment	0.10
Sodium citrate	Buffer + electrolyte	0.25
Curcumin emulsion (10%)	Functional antioxidant	0.05
Ginger extract (spray-dried)	Aroma and function	0.02
Ascorbic acid	Antioxidant stabilizer	0.03
Natural flavor (lemon–mint)	Sensory brightness	0.03
Erythritol	Sweetness	0.40
Gum Arabic	Stabilizer	0.05

Performance Characteristics:

- pH: 3.8
- Thermal tolerance: stable at 85 °C (30 s)
- Clarity: <1.2 NTU
- Shelf life: ≥12 months at 25 °C
- Functional retention: ≥90% curcuminoids, ≥85% ascorbate

19.5.9 Formulation Insight

- Treat stability as a system, not a property, pH, oxygen, temperature, and packaging are interconnected.

- Select extracts according to their stability class (e.g., anthocyanins for cold-fill; turmeric for hot-fill).
- Always validate through accelerated testing before scaling up.
- Sensory rebalancing post-storage is as important as chemical retention, freshness perception is part of shelf life.
- The most successful botanical beverages achieve functional efficacy and sensory integrity simultaneously, even after a year on the shelf.

20. Protein, Collagen, and Amino Acid Integration Systems in Hydration and Recovery Beverages

Protein fortification in hydration beverages is no longer limited to sports shakes; it is a growing frontier across active lifestyle, beauty-from-within, and functional wellness categories.
The challenge is to incorporate protein or amino acid systems that remain soluble, stable, and appealing in aqueous, often acidic, ready-to-drink (RTD) environments.

This chapter provides a framework for understanding protein chemistry, solubility behavior, stabilization approaches, and amino acid balance strategies.

20.1 Functional Role of Proteins and Amino Acids in Hydration and Recovery

Proteins and amino acids contribute essential physiological support during hydration and recovery by aiding muscle repair, reducing exercise-induced damage, and supporting metabolic pathways that restore cellular function. Their presence in hydration systems enhances nutrient delivery, stabilizes fluid balance, and provides substrates for rebuilding structural proteins and enzymes. When incorporated into beverages, these components elevate the formulation from simple replenishment to targeted recovery support.

Table 20.1: Functional Protein and Amino Acid Matrix for Hydration and Recovery Beverages

Functional Category	Typical Protein Source	Key Amino Acids	Beverage Type	Target Outcome
Recovery	Whey isolate, hydrolyzed collagen	BCAA (Leu, Ile, Val)	Post-exercise drinks	Muscle repair, rehydration

Hydration + Energy	Pea, rice, or soy hydrolysate	Glutamine, alanine	Isotonic recovery drinks	Electrolyte balance + metabolism
Beauty / Skin Support	Collagen peptides	Glycine, proline, hydroxyproline	Clear collagen waters	Skin elasticity, joint support
Cognitive / Nootropic	Amino acid blends	Tyrosine, theanine	Functional tonics	Focus, relaxation
Plant Protein Blends	Pea, rice, chia	Complete EAA profile	Vegan recovery drinks	Nutrient density, sustainability

Formulator's Insight:

Aim for 5–10 g protein per 500 mL serving for recovery or beauty-positioned drinks; higher levels (>15 g) typically require opaque bases or dairy systems.

20.2 Solubility and Stability Challenges of Protein Systems

Protein systems introduce several formulation challenges, including limited solubility, precipitation, and susceptibility to pH- and heat-induced denaturation. Whey, plant isolates, and collagen differ widely in their behavior under acidic, electrolyte-rich conditions typical of hydration products. Managing hydration temperature, pH, shear, and mineral load is essential to maintaining clarity, preventing sedimentation, and ensuring a smooth, stable sensory profile.

Protein stability in beverages is influenced by pH, ionic strength, temperature, and protein type.

Most proteins denature or precipitate outside their optimal pH or when exposed to heat during pasteurization.

Table 20.2: Protein Solubility and Stability Profile

Protein Type	Isoelectric Point (pI)	Solubility Range	Key Compatibility Notes
Whey Isolate	4.6–5.2	Best at pH 2.8–4.0 or >6.0	Excellent clarity below pH 4; sensitive to heat
Collagen Peptides	7.0–9.0	Fully soluble pH 3–7	Excellent heat and acid stability
Pea Protein Hydrolysate	4.5	Soluble pH >6	Requires enzymatic hydrolysis for clear systems
Soy Isolate	4.5	Soluble pH >6	For neutral pH beverages only
Rice Protein Hydrolysate	4.8	Moderate pH 6–7	Mild flavor, low viscosity

Best Practices:

- Collagen peptides are ideal for clear or acidic hydration beverages (stable up to 90 °C).
- Use protein hydrolysates to enhance solubility and minimize turbidity.
- Avoid combining divalent cations (Ca^{2+}, Mg^{2+}) with protein bases at low pH, they trigger precipitation.
- Add proteins after acidification to minimize denaturation during blending.

20.3 Collagen Systems in Beauty and Hydration Formulations

Collagen has emerged as a key ingredient in beauty-forward hydration beverages because of its perceived benefits for skin elasticity, joint health, and overall connective tissue support. Collagen peptides offer excellent solubility and neutral flavor, making them well-suited to clear, low-viscosity beverages. Their integration requires careful control of pH and flavor pairing, but when executed well, collagen systems provide a premium, functional differentiator for modern hydration formulations.

Collagen peptides (type I and III) are highly stable and neutral-tasting, making them excellent for beauty or functional hydration waters.

Key Properties:

- Soluble across pH 3–8.
- Heat-stable up to 90 °C (short time).
- Contribute minimal turbidity (<1.5 NTU).
- Compatible with natural flavors, botanicals, and polyols.
- Support health claims for skin elasticity, hydration, and joint comfort (subject to local regulations).

Formulation Guidelines:

- Use 2.5–10 g per serving depending on claim intensity.
- Maintain pH 3.8–4.2 for taste and clarity.
- Combine with vitamin C (0.03–0.05%) to enhance collagen synthesis.
- Stabilize with mild buffer (sodium citrate 0.25%).
- Optional addition: hyaluronic acid 0.01–0.02% for synergistic effect.

Example Positioning:

"Radiant Hydration Water", clear drink with collagen peptides, vitamin C, and rose–lychee flavor for skin vitality.

20.4 Managing Protein Interactions with Electrolytes and Acids

Proteins interact dynamically with electrolytes and acids, often leading to aggregation, charge imbalances, or structural changes that affect stability. Divalent minerals like calcium and magnesium are particularly reactive, while low-pH environments can reduce solubility or trigger precipitation. Successful formulation requires balancing acidity, mineral concentration, and protein type while leveraging stabilizers or buffering systems to maintain clarity and texture.

Electrolytes are essential in recovery formulations, but their ionic strength can destabilize protein systems.

Table 20.4: Protein–Electrolyte Compatibility Matrix

Ingredient Interaction	Risk	Stabilization Strategy
Whey protein + Ca^{2+}	High precipitation	Use sodium salts; chelate Ca^{2+} with citrate
Collagen + Na^+ / K^+	Low risk	Compatible up to 0.5% total salt
Plant protein + Acidic pH	Phase separation	Use hydrolyzed plant protein or encapsulated acids
Amino acids + Electrolytes	pH drift	Buffer to pH 3.8–4.0
Botanicals + Proteins	Astringency	Add small pectin or glycerol (0.05%) to smooth texture

Formulation Tip:

Introduce proteins post-electrolyte addition and under continuous stirring at <40 °C.

20.5 Amino Acid Fortification Strategies

Amino acid fortification allows developers to target specific recovery pathways, enhance hydration performance, and address the limitations of whole-protein systems. Branched-chain amino acids support muscle repair, while glutamine and alanine contribute to metabolic recovery and fluid balance. Incorporating amino acids in free form provides predictable solubility and pH stability, enabling precise dosing and creating opportunities for highly tailored functional profiles within hydration beverages.

Amino acids can enhance both functional performance and taste architecture. However, several free amino acids have bitterness or metallic notes that must be masked or balanced.

Table 20.5: Amino Acid Fortification Design

Amino Acid	Functional Role	Taste Profile	Masking Strategy
Leucine / Isoleucine / Valine (BCAA)	Muscle recovery	Bitter	Use fruit acids + vanillin
Glutamine	Cellular hydration, gut health	Mild	Naturally compatible
Alanine	Osmotic regulation	Slightly sweet	Enhances flavor roundness
Tyrosine	Mental focus	Bitter	Pair with citrus or herbal notes
Theanine	Calm focus	Umami	Works with green tea systems
Glycine	Collagen synthesis	Sweet	Acts as taste smoother

Taste Engineering Approach:

- Maintain total amino acid load ≤1% for clear RTDs.
- Combine with erythritol or glycine to reduce bitterness.
- Use citrus or floral notes to uplift heavy amino acid bases.

20.6 Example Formulation - Collagen + Electrolyte Hydration Drink

A collagen + electrolyte hydration drink demonstrates how collagen peptides, sodium and potassium salts, light organic acids, and subtle fruit flavors can combine to produce a clean, functional beverage. This formulation balances protein clarity, mineral compatibility, and sensory smoothness, delivering both hydration and beauty-oriented benefits. It highlights the practical considerations needed to integrate protein functionality into a clear, refreshing format.

Table 20.6: Collagen–Electrolyte Hydration Beverage

Ingredient	Function	% w/w
Purified water	Solvent	q.s. to 100
Collagen peptides	Functional protein	1.0
Citric acid	pH control	0.10
Sodium citrate	Buffer + electrolyte	0.25
Potassium chloride	Electrolyte	0.10
Vitamin C	Cofactor for collagen synthesis	0.03
Natural flavor (rose–lychee)	Sensory	0.05
Erythritol	Sweetness	0.40
Gum Arabic	Texture balance	0.05

Performance Metrics:

- pH: 3.8
- Clarity: <1.2 NTU
- Protein recovery: >95% post-fill
- Shelf life: 12 months (ambient)
- Sensory: floral, refreshing, slightly creamy mouthfeel

20.7 Formulation Insight

Insight emerges when protein behavior, mineral interactions, pH dynamics, and sensory constraints are evaluated together rather than independently.

Understanding how these variables shape solubility, stability, and function enables developers to choose the right protein sources, complementary amino acids, and stabilizing tools. This integrated approach leads to hydration beverages that deliver measurable recovery benefits while maintaining clarity, consistency, and modern sensory appeal.

- Protein and electrolyte systems must be chemically compatible, not just nutritionally complementary.
- Hydrolysis and pH control are your main tools for clear, stable protein beverages.
- Collagen remains the most forgiving and versatile protein for acidic, clear beverages.
- Use amino acid synergy to fine-tune performance claims, for instance, pairing collagen (structure) with glycine and vitamin C (synthesis).
- Always validate protein solubility and clarity under accelerated conditions (40 °C for 4 weeks).

21. Vitamin and Micronutrient Fortification Systems

Modern hydration and functional beverages often serve as delivery systems for essential vitamins and trace minerals, not just for rehydration but for energy metabolism, immune support, cognitive performance, and overall wellness. However, vitamins and minerals are chemically delicate, often prone to oxidation, pH drift, and light degradation.

The key to effective fortification is controlled stabilization and intelligent formulation architecture that ensures nutrient integrity over the entire product shelf life.

21.1 The Functional Role of Vitamins in Hydration and Recovery

Vitamins support core physiological processes that enhance hydration and recovery, influencing energy metabolism, antioxidant defense, neuromuscular function, and overall cellular resilience. Their inclusion transforms a simple hydration beverage into a performance tool, provided they are delivered in bioactive, stable forms. Understanding these functional roles sets the stage for designing beverages that contribute meaningfully to post-exercise restoration and daily wellness.

Each vitamin contributes to a distinct physiological function, and beverage developers typically target synergistic vitamin clusters that reinforce the brand's positioning (e.g., "Energy," "Focus," "Immunity," or "Beauty").

Table 21.1: Functional Vitamin Clusters for Hydration and Recovery Beverages

Functional Category	Key Vitamins	Core Function	Beverage Type

Hydration + Energy	B1, B2, B3, B5, B6, B12, C	Carbohydrate metabolism, fatigue reduction	Isotonic, electrolyte drinks
Immunity & Antioxidant	C, D, E, A, zinc, selenium	Antioxidant defense, immune modulation	Functional waters, cold tonics
Beauty / Skin Health	C, E, biotin, collagen support	Collagen synthesis, antioxidant	Collagen waters, wellness drinks
Cognitive & Focus	B6, B12, folate, choline	Neurotransmitter synthesis	Nootropic hydration beverages
Recovery & Performance	C, B-complex, D	Redox balance, energy regeneration	Protein or post-workout drinks

Formulator's Insight:

When pairing vitamins with electrolytes, maintain a balanced ionic and pH environment (3.5–4.2), both for stability and palatability.

21.2 Vitamin Stability Factors

Vitamin stability is shaped by a complex interplay of light exposure, oxygen, pH, heat, minerals, and ingredient interactions. Even small shifts in formulation conditions can lead to degradation, potency loss, or off-flavors. Recognizing which factors threaten stability allows developers to anticipate vulnerabilities and protect vitamin efficacy throughout processing and storage.

Vitamins differ widely in their chemical resilience.

Understanding their degradation mechanisms is fundamental for system design.

Table 21.2: Vitamin Stability Profile

Vitamin	Stability to Heat	Light	Oxygen	pH Sensitivity	Notes
Vitamin C (Ascorbic Acid)	Moderate	Low	Low	Unstable above pH 4.5	Use with chelators and nitrogen sparging
Vitamin B1 (Thiamine)	Low	Moderate	Moderate	Unstable < pH 3	Prefer thiamine mononitrate in acid drinks
Vitamin B2 (Riboflavin)	High	Very Low	Moderate	Neutral	Protect from light (amber PET)
Vitamin B3 (Niacinamide)	High	High	High	Stable	Ideal for acidic drinks
Vitamin B6 (Pyridoxine)	Moderate	Moderate	Moderate	Stable	Compatible with electrolytes
Vitamin B12 (Cyanocobalamin)	Low	Very Low	Low	Sensitive < pH 3	Use microencapsulated B12
Vitamin D3	Moderate	Low	Low	Stable	Add post-pasteurization
Vitamin E (Tocopherol)	High	Low	Low	Stable	Protect with emulsifiers (oil dispersion)
Folic Acid	Low	Moderate	Low	Unstable in light + heat	Add cold; protect with encapsulation

Biotin (H-B7)	High	High	High	Stable	Use as soluble form for clear beverages

Critical Note:

Most B vitamins are prone to color shifts and off-flavors under light exposure, always use UV-protected packaging.

21.3 Stabilization and Delivery Strategies

Effective stabilization relies on choosing the right vitamin forms, protective carriers, encapsulation systems, antioxidants, and processing conditions. These strategies help preserve potency and ensure consistent delivery across the product's shelf life. By aligning stabilization tools with formulation demands, developers can maintain nutritional integrity without compromising clarity, flavor, or texture.

1. Encapsulation and Coating Technologies

Encapsulation protects labile vitamins from moisture, oxygen, and reactive beverage components.

Table 21.3: Vitamin Encapsulation and Delivery Technologies

Technology	Description	Best For	Example Application
Spray-Dried Matrix	Vitamins embedded in maltodextrin or gum arabic	C, B-complex	Powder premixes, dry-fill drinks
Microencapsulation (Lipid)	Vitamin droplets enclosed in fat or starch shell	D, E, K, folic acid	RTD and dairy-like beverages
Cyclodextrin Inclusion	Forms molecular cage around actives	Fat-soluble vitamins	Transparent drinks

Coated Granules	Polymer or protein-coated vitamins	B-complex blends	Effervescent or long-shelf products

Formulation Tip:

Use multi-layer encapsulation (hydrophilic + lipid) for RTDs with mixed vitamin classes.

2. Chelation and Antioxidant Coupling

- Citric acid (0.1–0.2%) and sodium citrate (0.25%) chelate trace metals, reducing oxidative degradation.
- Ascorbic acid + tocopherol systems regenerate each other in solution (vitamin E regenerates oxidized C).
- Use glucose oxidase or oxygen absorbers in caps for ultra-sensitive vitamins like B12.

3. pH and Temperature Control

- Maintain pH 3.5–4.0 to balance microbial safety and vitamin stability.
- Minimize thermal exposure: use cold-fill or aseptic blending for heat-labile vitamins.
- If hot-fill is necessary, over-fortify heat-sensitive vitamins by 10–15% to account for process loss.

21.4 Mineral Fortification and Compatibility

Fortifying beverages with minerals introduces challenges in solubility, reactivity, flavor impact, and compatibility with vitamins. Ionic strength and pH can influence both nutrient stability and sensory outcome. Understanding how minerals interact within the broader system enables formulators to achieve balanced nutrient delivery while avoiding precipitation, discoloration, and off-notes.

Micronutrient minerals (zinc, magnesium, calcium, iron, selenium) enhance performance and wellness but can cause instability, off-tastes, or interactions with vitamins and botanicals.

Table 21.4: Mineral Fortification Compatibility Matrix

Mineral	Preferred Form	Compatibility	Notes
Calcium	Calcium lactate or gluconate	Limited (precipitates with phosphates)	Avoid >200 ppm in clear drinks
Magnesium	Magnesium citrate or lactate	Good	Smooth taste; aids metabolism
Zinc	Zinc gluconate	Moderate	Avoid >10 ppm (metallic taste)
Iron	Ferrous bisglycinate (chelated)	Fair	Use only in opaque formulations
Selenium	Sodium selenite	High	Stable under acidic pH
Chromium	Chromium picolinate	High	Stable; used in energy metabolism blends

Best Practices:

- Prefer organic salts (lactate, citrate, gluconate) for taste and solubility.
- Always validate color and pH stability in the presence of polyphenols or proteins.
- For clear drinks, limit total mineral concentration to ≤0.5% TDS to maintain visual clarity.

21.5 Premix Engineering and Quality Control

Premix engineering ensures that vitamins and minerals are blended in forms that remain stable, flowable, and consistently dosed during production. Quality control, through assay verification, particle size management, and carrier selection, safeguards nutritional accuracy and regulatory compliance. Strong premix design reduces risk and streamlines manufacturing across multiple batch sizes and processing environments.

Vitamin-mineral premixes are often custom-built for production consistency. Key considerations include particle size, flowability, hygroscopicity, and homogeneity.

Premix Design Checklist:

1. Ensure uniform blending, add flow agents (silica ≤1%).
2. Use low-moisture carriers (maltodextrin, microcrystalline cellulose).
3. Validate assay recovery post-filling (≥90% of label claim).
4. Include antioxidants if C or E are part of the blend.
5. Maintain storage RH <40% and temperature <25 °C.

Analytical Verification:

- HPLC / UV–Vis: Vitamin quantification.
- ICP–MS: Trace mineral analysis.
- Accelerated testing: 40 °C / 75% RH, 3 months.

21.6 Example Formulation - Electrolyte + Vitamin Performance Drink

An electrolyte + vitamin performance drink illustrates how hydration minerals, stabilizers, vitamin forms, and flavor systems can be integrated into a cohesive, high-functioning formulation. The example highlights compatibility considerations, stability strategies, and sensory balancing techniques that support both performance benefits and consumer appeal.

Table 21.6: Electrolyte + Vitamin Performance Drink - B-Complex + Vitamin C Synergy, pH 3.8, NTU <1.0, ≥90% Vitamin Retention (6 Months)

Ingredient	Function	% w/w
Purified water	Solvent	q.s. to 100
Citric acid	pH control	0.10
Sodium citrate	Buffer + electrolyte	0.25
Potassium chloride	Electrolyte	0.10
Vitamin C (ascorbic acid)	Antioxidant	0.03
Niacinamide (B3)	Energy metabolism	0.02
Pyridoxine HCl (B6)	Coenzyme support	0.01
Cyanocobalamin (B12, encapsulated)	Cellular energy	0.0005
Natural lemon flavor	Sensory	0.05
Erythritol	Sweetness	0.40
Gum Arabic	Stabilizer	0.05

Performance Metrics:

- pH: 3.8
- Clarity: <1.0 NTU
- Vitamin retention: >90% after 6 months at 25 °C
- Shelf life: 12 months (light-protected PET)

21.7 Formulation Insight

Insight emerges when vitamin behavior, mineral interactions, stabilizer selection, and processing constraints are viewed as interconnected variables. Identifying the factors most responsible for degradation or sensory drift helps developers refine their systems with precision. These learnings guide the creation of beverages that deliver reliable nutrition, strong shelf stability, and a clean, consistent sensory experience.

- Design vitamin systems with chemical realism, assume oxygen, light, and time will attack them, and build your shield accordingly.
- Always match processing type to vitamin stability class (heat-labile vs. heat-tolerant).
- Favor encapsulation, antioxidants, and chelation over simply over-fortifying.
- Transparency and clean label expectations require natural carriers and gentle processes.
- Vitamin and mineral stability defines consumer trust, efficacy must last as long as the beverage's shelf life.

22. Flavor Masking, Taste Balancing, and Sensory Engineering Systems

Functional hydration beverages often contain ingredients that challenge taste perception: salty electrolytes, bitter amino acids, metallic minerals, or earthy botanicals.
The difference between a "healthy drink" and a "habit-forming beverage" is not merely sweetness or acidity, it is the precision of sensory balance, achieved through layering, masking, and modulation systems.

This chapter explores the scientific foundations and practical techniques for taste correction, flavor enhancement, and multi-sensory harmony in modern electrolyte and nutrition beverages.

22.1 The Science of Taste Interactions

Taste interactions define how the palate perceives sweetness, acidity, bitterness, and functional off-notes, especially in beverages containing electrolytes, amino acids, botanicals, or proteins. Natural and functional flavor masking leverages these interactions to soften harsh edges, create balance, and support a clean overall profile. Understanding these mechanisms provides the basis for designing beverages that deliver both performance and palatability.

Functional ingredients often distort these balances.

Table 22.1: Taste Modality Interactions in Functional Hydration Beverages

Taste Note	Source in Formulations	Sensory Challenge	Desired Counterbalance
Salty	Sodium, potassium, magnesium salts	Harshness, mineral aftertaste	Sweet–acid pairing, fruit acids
Bitter	Polyphenols, BCAAs, caffeine	Lingering harshness	Natural sweeteners, vanilla, masking agents
Astringent	Polyphenols, tannins	Dry mouthfeel	Glycerol, pectin, low-acyl gellan gum

Metallic	Zinc, iron, copper	Off-flavor	Citric acid chelation, maltol, caramel notes
Sulfuric / earthy	Adaptogens, amino acids	Musty perception	Citrus top notes, mint, floral bridges
Sour / acidic	Citric, malic, ascorbic acids	Sharp or puckering	Low-intensity sweeteners, sodium balance

Formulator's Insight:

Taste balancing is not linear. A 10% reduction in acidity may require a 40% increase in perceived sweetness to restore equilibrium, always validate via sensory panels.

22.2 Natural and Functional Flavor Masking Approaches

Flavor masking can be achieved through multiple approaches, modulating pH, shaping sweetness curves, layering complementary aromas, or deploying targeted masking agents. Each method addresses different sensory challenges and interacts uniquely with electrolytes and functional ingredients. Selecting the right approach ensures that flavor complexity is managed without compromising stability or nutritional intent.

Table 22.2: Clean-Label Flavor Masking Architecture

Masking Strategy	Mechanism	Example Ingredients	Application Notes
Sweet–Bitter Masking	Sweetness suppresses bitter receptor signaling	Stevia + erythritol + vanilla	Ideal for BCAA or polyphenol drinks
Aromatic Diversion	Aroma distracts taste focus	Citrus, mint, spice top-notes	High volatility essential oils

Acid–Base Contrast	Balances pH perception and sourness	Citric + malic acid system	Enhances fruit brightness
Mouthfeel Modulation	Adds body and smoothness to offset bitterness	Glycerol, pectin, gum arabic	For thin, clear drinks
Thermal Sensory Trick	Cooling or warming effect masks sharp notes	Menthol, capsaicin analogs	Used in mint–ginger functional blends
Flavor Fusion	Bitter + compatible note forms "new pleasant profile"	Cocoa–ashwagandha, tea–citrus	Strategic formulation choice

Practical Tip:

Combine two or more sensory tools, for example, a mild acid–sweet system (citric + erythritol) with vanilla–mint aroma, to fully neutralize complex off-notes.

22.3 Balancing Sweetness and Acidity for Functional Harmony

The interplay between sweetness and acidity forms the backbone of flavor harmony, influencing perceived freshness, masking efficiency, and how well functional ingredients integrate into the profile. Fine-tuning this balance helps neutralize bitterness, enhance fruit notes, and support overall drinkability. Mastery of sweetness–acidity dynamics enables developers to create hydration and recovery drinks that taste both clean and purposeful.

Sweetness–Acidity Balance Zones (Typical pH 3.6–4.0):

Table 22.3: Sweetness–Acidity Balance Matrix for Hydration Beverage Categories

Beverage Type	Sweetness Intensity (Sucrose Equivalent, %)	Acidity (citric acid %)	Sensory Goal
Isotonic Hydration	3.5–5.0	0.08–0.12	Refreshing, clean
Low-Calorie Functional Water	1.5–2.5	0.05–0.08	Light, crisp
Botanical Antioxidant Drink	2.5–3.5	0.08–0.10	Balanced, soft tartness
Collagen Beauty Drink	2.0–3.0	0.05–0.09	Delicate, floral-sweet
BCAA / Protein Drink	4.0–5.5	0.10–0.15	Smooth, no aftertaste

Rule of Thumb:

Acidity should represent about 2–3% of perceived sweetness intensity, beyond this, taste fatigue and throat burn increase.

22.4 Advanced Flavor Masking Technologies

Advanced technologies, such as microencapsulation, delayed-release flavor systems, bitterness blockers, and selective botanical fractions, offer precise tools for managing difficult off-notes. These systems modify how flavors are released, perceived, and interpreted by the palate, often improving stability under heat, acid, and mineral stress. Leveraging these technologies allows for more sophisticated, high-performance flavor solutions.

Table 22.4: Advanced Taste Modulation Technologies - Receptor-Level Interventions for Bitterness Suppression and Sweetness Enhancement in Functional Beverages

Technology	Mechanism	Example	Application

Receptor Blockers	Temporarily block bitter receptor sites	Lactisole, thaumatin, adenosine analogs	Caffeine, polyphenol drinks
Sweetness Modulators	Enhance perception of sweetness	Mogrosides, monk fruit, glycyrrhizin	Low-calorie beverages
Flavor Encapsulation	Delayed release masks initial bitterness	β-Cyclodextrin inclusion	BCAA, caffeine systems
Lipid-Coated Bitterants	Isolates taste receptors from contact	Microencapsulated amino acids	Functional waters
Cross-Modal Enhancement	Leverages aroma to "amplify" taste	Vanilla, berry, citrus	Clean-label flavor synergy

Formulator's Insight:

Microencapsulation provides a double advantage: it reduces immediate bitterness and enhances stability of active compounds (e.g., caffeine, theanine, curcumin).

22.5 Mouthfeel and Texture Enhancement

Mouthfeel contributes to overall flavor perception by shaping fullness, smoothness, and the tactile qualities that support masking performance. Gums, fibers, and carbohydrate systems can soften sharp notes, reduce astringency, and create a more integrated sensory experience. Understanding the role of texture helps formulators create beverages that feel balanced, cohesive, and satisfying.

Preferred Texturizing Agents for Clear Functional Drinks:

Table 22.5: Preferred Texturizing Agents for Clear Functional Drinks: Function, Use Level, and Sensory Effect

Ingredient	Function	Use Level (%)	Sensory Effect
Glycerol	Texture and sweetness booster	0.2–0.5	Softens acidity
Pectin (low methoxyl)	Mouthfeel enhancement	0.05–0.10	Smooth and natural
Gum Arabic	Body and flavor fixative	0.05–0.10	Rounds off sharp notes
Gellan gum (low-acyl)	Suspension + clarity	0.02–0.04	Crisp, clean texture
Polydextrose	Low-calorie bulking agent	0.3–0.6	Adds weight, balances electrolytes

Formulation Tip:

For clear beverages, avoid gums that cloud (e.g., guar, xanthan); instead, select gellan or gum arabic for invisible textural support.

22.6 Aroma Architecture and Psychological Masking

Aroma plays a powerful psychological role in masking, guiding the brain toward expected flavors and away from functional off-notes. Structuring aroma architecture through top notes, mid-tones, and finishes can reshape perception and elevate the entire sensory profile. Harnessing these cues enables beverages to taste cleaner and more refined while still delivering high functional loads.

Aromas can redirect attention from undesirable taste notes by stimulating memory and expectation.
Aromas that cue "freshness," "cleanliness," or "naturalness" psychologically mask functional harshness.

Table 22.6: Aroma-Driven Psychological Masking Framework

Aroma Family	Perceptual Effect	Ideal Applications
Citrus (lemon, yuzu, bergamot)	Bright, cleansing	Electrolyte waters
Mint / Menthol	Cooling, refreshing	Adaptogen or caffeine systems
Floral (jasmine, rose, hibiscus)	Elegant, soft	Beauty / collagen drinks
Spice (ginger, cardamom)	Warm, functional	Recovery tonics
Green / Herbal (matcha, basil)	Natural, earthy	Botanical waters
Berry / Tropical (acai, mango, lychee)	Fruity, energizing	Antioxidant and immune blends

Practical Example:

A bitter ashwagandha base becomes pleasant when paired with rose–vanilla top notes and soft acidity (pH 3.8), creating a "floral chai" impression rather than an herbal tonic.

22.7 Example Formulation - BCAA Citrus–Vanilla Recovery Water

A BCAA citrus–vanilla recovery water showcases how layered acidity, controlled sweetness, aroma design, and functional masking tools can work together to smooth the inherent bitterness of amino acids. This formulation illustrates how flavor chemistry, texture modulation, and aroma cues align to create a refreshing, performance-oriented recovery drink.

Table 22.7: BCAA Citrus–Vanilla Recovery Water

Ingredient	Function	% w/w
Purified water	Solvent	q.s. to 100

Citric acid	pH control	0.10
Sodium citrate	Buffer + electrolyte	0.25
Potassium chloride	Electrolyte	0.10
BCAA blend (2:1:1)	Functional actives	0.20
Natural vanilla flavor	Base note	0.03
Natural lemon oil (emulsified)	Top note	0.05
Erythritol	Sweetness	0.40
Stevia (Reb M)	Sweetener	0.015
Glycerol	Mouthfeel	0.25
Gum Arabic	Texture + stabilization	0.05

Performance Metrics:

- pH: 3.8
- Clarity: <1.0 NTU
- Taste: citrus-forward with soft vanilla finish
- Bitterness reduction: ~60% sensory suppression vs. unmasked control
- Shelf life: 12 months (cold-fill, nitrogen flushed)

22.8 Formulation Insight

Insight emerges when masking strategies, flavor balance, texture, and aroma design are treated as interconnected elements rather than independent solutions. Recognizing how each variable influences perception enables developers to refine flavor systems with precision and consistency. These learnings help create beverages that deliver functional benefits while maintaining a clean, enjoyable sensory experience.

- Masking should not neutralize; it should re-contextualize, transform a negative sensory cue into a functional signature.

- The most effective taste balance comes from sweet–acid–aroma triangulation, not simply from added sugar or flavor.
- Always run triangle sensory tests ($n \geq 10$ panelists) to validate masking effectiveness under real beverage conditions.
- Clean-label flavor masking requires smart chemistry, not tricking the palate, but guiding it.

23. Natural Color Systems and Stability Management

Color is the first sensory signal consumers register in a beverage. It conveys freshness, function, and even flavor expectation, a golden-yellow drink suggests citrus vitality; a pale pink, floral calm.
In functional and electrolyte beverages, color must do more than attract attention, it must remain stable across pH shifts, light exposure, and temperature variations.

Synthetic dyes offer simplicity, but natural colorants express authenticity, sustainability, and health.
This chapter outlines how to select, stabilize, and preserve natural pigments in functional drinks without compromising clarity or shelf life.

23.1 Functional Role of Color in Hydration and Nutrition Beverages

Color acts as an immediate sensory cue that shapes expectations around flavor, purity, and functional positioning. In hydration and nutrition beverages, color also communicates ingredient identity and perceived efficacy, influencing how consumers judge freshness and performance. Understanding its functional role allows developers to design visual profiles that support both brand intent and sensory experience.

Color signals both identity and quality.

For functional beverages, it also reinforces the product's emotional message (e.g., green = "clean energy," purple = "antioxidant," gold = "revitalizing").

Table 23.1: Functional Color Mapping: Hue, Source, and Psychological Positioning in Hydration Beverages

Functional Category	Desired Hue	Common Natural Source	Functional Message

Hydration / Citrus	Pale yellow–green	β-carotene, turmeric, chlorophyllin	Fresh, energizing
Recovery / Protein	Cream–gold	Carotenoid + vanilla tone	Comfort, warmth
Beauty / Collagen	Pink–rose	Anthocyanins, beet extract	Soft, radiant
Antioxidant / Botanical	Red–purple	Hibiscus, elderberry, grape skin	Potent, natural
Focus / Adaptogenic	Amber–tea	Curcumin, rooibos	Calm, mindful

Formulator's Insight:

The psychological connection between color and function strengthens brand recall. Always align pigment choice with both flavor direction and health positioning.

23.2 Major Classes of Natural Colorants

Natural colorants, from anthocyanins and carotenoids to betalains and chlorophylls, offer vibrant hues with unique chemical behaviors. Each class carries distinct stability patterns, solubility characteristics, and sensitivities to pH, light, and minerals. Mapping these differences helps formulators select colorants that align with the beverage's acidity, electrolyte load, and intended shelf-life conditions.

Table 23.2: Natural Colorant Classification by Chemistry, Stability, and Beverage Suitability

Colorant Class	Common Source	Hue Range	pH Sensitivity	Light / Heat Stability	Key Formulation Note

Anthocyanins	Berries, hibiscus, purple corn	Red → Blue	High (color shifts with pH)	Moderate	Best below pH 4
Betalains	Beetroot	Pink → Red	Low (stable across pH 3–7)	Low (heat-sensitive)	Cold-fill only
Carotenoids	Carrot, paprika, annatto	Yellow → Orange	Low	High (in emulsion)	Requires emulsification
Curcuminoids	Turmeric	Yellow → Orange	Low	High	Use encapsulated form
Chlorophyll / Chlorophyllin	Alfalfa, spinach	Green	Moderate	Moderate (light-sensitive)	Use copper-chlorophyllin for clarity
Spirulina Phycocyanin	Blue-green algae	Blue	Very high	Poor (heat/pH sensitive)	Use only in cold beverages

Color Selection Tip:

For acidic functional beverages (pH 3.5–4.2), anthocyanins, curcumin, and carotenoids are the most stable and visually appealing.

23.3 pH-Dependent Color Behavior

Many natural pigments shift hue dramatically with changes in acidity, creating opportunities for intentional color design but also risks of unintended visual drift. These pH-dependent transitions influence shade, intensity, and stability throughout production and storage. Understanding these behaviors allows developers to control color expression and maintain consistency across real-world conditions.

Colorants like anthocyanins are pH indicators, their hue shifts dramatically across the beverage pH range.

Table 23.3: Anthocyanin Chromatic Shifts Across Beverage pH Ranges

pH	Anthocyanin Appearance	Example Use
2.5–3.2	Bright red	Hibiscus–berry waters
3.5–4.0	Deep magenta	Collagen beauty drinks
4.5–5.0	Violet	Herbal tonics
>6.0	Blue–gray (unstable)	Avoid for beverages

Practical Strategy:

Always finalize flavor and acid composition first, then fine-tune colorant concentration under the finished pH, never the reverse.

23.4 Stabilization Mechanisms for Natural Colors

Stability challenges, oxidation, light sensitivity, and interactions with minerals, require targeted strategies to preserve natural pigments over time. Encapsulation systems, antioxidants, protective emulsions, and controlled pH environments all play critical roles in protecting hue and clarity. Applying these mechanisms ensures that natural colors remain vibrant and reliable throughout shelf life.

Color degradation arises from oxidation, light exposure, and metal-catalyzed reactions. Stability must be managed chemically and physically.

Table 23.4: Integrated Stabilization Strategies for Natural Pigments

Stabilization Strategy	Mechanism	Example Additives	Typical Level

pH Control	Maintains molecular form	Citric acid, malic acid	0.05–0.1%
Chelation	Removes metal catalysts	Sodium citrate, EDTA (natural alt: phytic acid)	0.1–0.2%
Antioxidant Protection	Slows pigment oxidation	Ascorbic acid, tocopherol	0.03–0.05%
Encapsulation / Emulsion	Protects from oxygen and light	Gum arabic, modified starch, β-cyclodextrin	1–3:1 carrier-to-color ratio
UV Shielding	Blocks photodegradation	Amber PET, opaque sleeve, aluminum can	Packaging choice

Formulator's Insight:

Curcumin and carotenoid systems perform best when co-encapsulated with antioxidants, while anthocyanins prefer low pH + low oxygen.

23.5 Clarity, Precipitation, and Light Scattering

Colorant solubility and particle behavior directly influence clarity and visual purity. Precipitation, flocculation, and light scattering can quickly undermine a beverage's appearance if color systems are not properly stabilized. Recognizing the physical factors that shape clarity enables formulators to achieve transparent, uniform beverages even when using complex natural pigments.

Even water-soluble colorants can cause haze if concentration, ionic strength, or particle size distribution are not optimized.

Table 23.5: Root Causes and Corrective Actions for Color-Related Physical Instability

Issue	Cause	Mitigation

Haze Formation	Aggregation of color particles or metal reaction	Fine filtration (0.45 μm), chelation
Precipitation	pH drift or salt interaction	Buffer system, avoid high calcium
Color Fading	Light and oxidation	Opaque packaging, antioxidant coupling
Uneven Color	Poor dispersion	Pre-hydrate colorant, use high-shear mixing

Clarity Tip:

For clear beverages, target NTU < 1.5, pre-emulsify any oil-soluble colorants before addition.

23.6 Example Color Systems and Their Application

Practical color systems, such as stabilized anthocyanin blends, oil-dispersed carotenoids, and micro-emulsified natural hues, demonstrate how pigment chemistry and formulation design intersect. Each system comes with defined performance ranges and ideal use-cases, providing developers with real-world models for achieving consistent coloration across diverse beverage styles.

Table 23.6: Application-Oriented Natural Color Systems and Commercial Use Cases

Functional Color System	Source	Target Hue	Beverage Example
Anthocyanin Complex	Hibiscus extract + elderberry	Ruby red	Antioxidant hydration
Carotenoid Emulsion	β-carotene in gum arabic	Sunny yellow	Citrus isotonic

Curcumin Encapsulate	Turmeric + starch matrix	Golden amber	Energy or recovery water
Copper-Chlorophyllin	Plant-based	Bright green	Detox / adaptogenic drink
Beet–Berry Blend	Betalain + anthocyanin mix	Soft pink	Beauty / collagen water

Formulator's Insight:

Dual-source blending (e.g., anthocyanin + carotenoid) widens color vibrancy and masks slight fading over time.

23.7 Example Formulation - Rose–Berry Collagen Water (Natural Red Hue)

A rose–berry collagen water showcases how anthocyanin-rich pigments, acidity control, and antioxidant protection can be combined to deliver a delicate natural red hue. This example highlights ingredient interplay, stability considerations, and sensory alignment, offering a template for visually refined, function-forward beverage development.

Table 23.7: Benchmark Formulation: Rose–Berry Collagen Water with Anthocyanin-Based Hue

Ingredient	Function	% w/w
Purified water	Solvent	q.s. to 100
Citric acid	pH control (3.8)	0.10
Sodium citrate	Buffer	0.25
Collagen peptides	Functional protein	1.0
Anthocyanin extract (hibiscus)	Natural color	0.015
Natural rose–lychee flavor	Sensory	0.05

Vitamin C	Antioxidant / cofactor	0.03
Erythritol	Sweetness	0.40
Gum Arabic	Stabilizer	0.05

Performance Metrics:

- pH: 3.8
- Color stability: >90% intensity after 6 months at 25 °C
- Clarity: <1.2 NTU
- Sensory: delicate floral, mildly tart berry-rose
- Shelf life: 12 months (light-protected PET)

23.8 Packaging Considerations for Color Retention

Packaging decisions, from barrier properties and light filtration to headspace design, play a critical role in maintaining natural color stability. The interaction between pigment sensitivity and packaging performance determines how well hues endure during distribution and storage. Aligning packaging with colorant vulnerabilities ensures long-term visual integrity.

Table 23.8: Packaging–Color Compatibility Matrix for Light-Sensitive Pigments

Packaging Type	Light Protection	Recommended Use
Amber PET (multi-layer)	Excellent	General functional beverages
Aluminum Can	Total protection	Long shelf-life isotonic drinks
Aseptic Carton	Good	Nutrient-enriched or dairy-based systems
Glass (dark)	Excellent but fragile	Premium beauty or botanical beverages

Design Note:

Avoid clear PET for anthocyanin-based drinks, even short UV exposure can cause noticeable fading.

23.9 Formulation Insight

Key insights emerge when color stability, clarity, pH management, and packaging design are treated as interconnected variables rather than isolated choices. Understanding these relationships empowers developers to refine color systems with precision, resulting in beverages that remain visually appealing, chemically stable, and aligned with functional and sensory goals.

- Treat color as a functional ingredient, not a decorative one, its chemistry defines stability.
- Always formulate colors under the final beverage pH, not laboratory pH water.
- Protect pigments with tri-layer stability systems: pH + antioxidant + packaging.
- Align visual tone with brand message and health function; color psychology is part of formulation strategy.
- Stability validation must include accelerated light and heat exposure tests (e.g., 1000 lux, 40 °C for 14 days).

24. Functional Sweetening and Carbohydrate Energy Systems

Sweetness is not only a flavor dimension, it is a functional architecture element that shapes mouthfeel, osmolarity, metabolic response, and consumer acceptance. In electrolyte and nutrition beverages, sweetness calibration determines whether the drink feels refreshing or cloying, energizing or heavy, and even premium or generic.

This chapter explores how to design sweetening systems that deliver ideal taste, metabolic function, and clean-label positioning, while maintaining isotonic balance and stability.

24.1 Functional Role of Carbohydrates and Sweeteners

Carbohydrates and sweeteners shape both the physiological function and the sensory experience of hydration beverages. Their roles extend beyond flavor, influencing energy delivery, fluid absorption, osmolality, and mouthfeel. Understanding how these components interact with electrolytes and acids provides the foundation for designing systems that balance taste with performance.

Carbohydrates influence three critical parameters in beverage design:

1. Osmolarity and Hydration Efficiency
    - Determines how fast water and electrolytes are absorbed.
    - Ideal osmolarity for isotonic beverages: 270–330 mOsm/kg.
2. Energy Delivery
    - Carbs provide quick or sustained energy depending on molecular weight.
    - Must align with the beverage's performance claim (e.g., "fast hydration" vs. "sustained endurance").
3. Taste and Mouthfeel
    - Sweetness perception balances electrolytes and acids.

- The right sugar-alcohol or rare sugar blend ensures roundness without stickiness.

24.2 Common Sweetening Options and Functional Profiles

Different sweetening agents, ranging from simple sugars and polyols to high-intensity sweeteners and natural extracts, carry distinct sweetness curves, stability behaviors, and metabolic implications. Each option influences flavor release, formulation cost, and hydration efficiency. Mapping their functional profiles allows developers to choose sweeteners that align with both product positioning and performance goals.

Table 24.2: Sweetener Systems: Functional Profiles, Glycemic Impact, and Clean-Label Suitability

Sweetener Type	Example	Relative Sweetness (Sucrose = 1.0)	Glycemic Impact	Key Attributes	Notes
Glucose (Dextrose)	Corn, tapioca	0.7–0.8	High	Fast energy	Excellent osmotic activity
Sucrose	Cane sugar	1.0	High	Balanced sweetness	Benchmark for taste calibration
Fructose	Fruit sugar	1.2–1.4	Moderate	Strong sweetness at low dose	Increases osmolarity quickly
Maltodextrin (DE 10–20)	Corn, rice	0.2–0.4	Medium	Sustained energy	Good for powder formulations
Isomaltulose (Palatinose)	Enzymatic sucrose	0.4–0.5	Low	Slow-release carbohydrate	Ideal for functional hydration
Allulose	Rare sugar	0.7	Low	Natural sweet taste	Minimal calories,

					excellent synergy
Erythritol	Sugar alcohol	0.6–0.7	None	Cooling effect, 0 kcal	Non-hygroscopic, clean label
Stevia (Reb A / Reb M)	Stevia leaf	200–300	None	High-intensity natural sweetener	Use in combination to mask aftertaste
Monk Fruit (Mogrosides)	Luo Han Guo	150–200	None	Natural sweet taste	Excellent for premium clean-label drinks
Agave / Honey Solids	Natural blends	1.1–1.3	Moderate	Natural positioning	Contribute flavor tone

Formulator's Insight:

Combine low-glycemic carbohydrates (isomaltulose, allulose) with high-intensity sweeteners (stevia, monk fruit) to achieve target sweetness while maintaining osmotic balance and metabolic neutrality.

24.3 Designing Sweetness Architecture

Sweetness architecture involves constructing a layered perception of sweetness that aligns with the beverage's flavor system, carbohydrate content, and intended use-case. Timing, onset, peak intensity, and finish all shape how consumers experience the drink. Thoughtful design ensures that sweetness supports flavor clarity, masks electrolyte harshness, and delivers a harmonious, balanced profile.

Sweetness should never exist in isolation, it must balance acidity, electrolytes, and flavor volatility.

Sweetness Curve Engineering:

- Onset: Fast sweetness (fructose, erythritol) masks initial saltiness.
- Mid-Persistence: Sucrose-equivalent or allulose provides roundness.
- Finish: Stevia or monk fruit extends sweetness tail and offsets bitter aftertaste.

Example 3-Layer Sweetness System:

Table 24.3: Sweetness Architecture Framework: Onset, Body, and Finish Engineering

Function	Ingredient	Use Level (%)
Fast-onset	Fructose or erythritol	0.2–0.4
Mid-body	Allulose or isomaltulose	0.3–0.6
Long-tail	Stevia Reb M	0.01–0.015

Sweet–Acid Ratio Guide (for pH 3.6–4.0):

- Balanced hydration: Sweet:Acid = 40:1
- Fruit-forward beverage: Sweet:Acid = 25:1
- Botanical tonic: Sweet:Acid = 60:1

24.4 Managing Osmolarity and Hydration Efficiency

Sweeteners and carbohydrates directly influence osmolality, a key determinant of fluid absorption and hydration performance. Managing these components requires balancing energy delivery with the need for rapid uptake, especially in isotonic and hypotonic formulations. Understanding how various sugars and osmolytes contribute to solution concentration enables developers to engineer beverages that hydrate efficiently without compromising function.

Too low → poor absorption (hypotonic).

Too high → slows gastric emptying (hypertonic).

Table 24.4: Osmolarity–Carbohydrate Design Matrix for Hydration Efficiency

Beverage Type	Target Osmolarity (mOsm/kg)	Carbohydrate Level (%)	Functional Intent
Hypotonic	150–250	1–2	Rapid rehydration

Isotonic	270–330	3–8	Balanced hydration
Hypertonic	350–500	8–15	Energy replenishment

Formulation Tip:

Use isomaltulose or maltodextrin (DE <15) to increase carbohydrate content without excessive osmolarity rise, ideal for endurance recovery drinks.

24.5 Clean-Label and Natural Sweetener Strategies

Clean-label expectations are reshaping sweetener choices, pushing formulators toward natural sugars, plant-derived extracts, and minimally processed options. These systems offer strong consumer appeal but come with challenges in stability, flavor balance, and interaction with electrolytes or acids. Strategic use of natural sweeteners allows developers to meet transparency demands while maintaining performance and sensory quality.

Table 24.5: Clean-Label Sweetening Strategies and Natural Synergy Systems

Natural System	Components	Functional Advantage	Typical Blend
Plant-Derived Hybrid	Stevia + erythritol	Smooth sweetness, 0 kcal	1:20 ratio
Botanical Infused Sweetness	Monk fruit + herbal extract	Perceived "natural health"	1:1–1:2 ratio
Low-Glycemic Base	Isomaltulose + allulose	Sustained energy	1:1 ratio
Honey or Agave Accent	Trace honey solids (<0.2%)	Rounds acidity, adds aroma	In beauty or recovery drinks

Practical Note:

Always evaluate sweetener synergy under final beverage pH, stevia's sweetness intensity rises at lower pH, while erythritol's perception diminishes.

24.6 Managing Sweetener Stability and Flavor Compatibility

Sweetener stability is influenced by pH, temperature, mineral content, and storage conditions, factors that can shift sweetness intensity or create off-notes over time. Effective management requires selecting sweeteners that remain consistent across processing and distribution while supporting the beverage's flavor architecture. Aligning stability with flavor compatibility ensures a clean, predictable sweetness profile through the product's shelf life.

Key Stability Factors:

Table 24.6: Sweetener Stability Under Processing and Formulation Stress

Stress Condition	Sensitive Ingredients	Solution
High Temperature (Hot-fill)	Allulose (caramelization), Stevia (flavor drift)	Use cold-fill or aseptic blending
Low pH (<3.2)	Erythritol (crystallization), Maltodextrin (hydrolysis)	Buffer to pH 3.5–3.8
Electrolyte Interaction	Stevia + NaCl = bitterness amplification	Pair with vanilla or maltol
Botanical Polyphenols	Can suppress sweetness perception	Add glycerol or increase top-note aroma

Stability Rule:

If sweeteners or carbohydrates exceed 3% total solids, validate osmolarity and sweetness perception every 3 months under accelerated testing (40 °C, 75% RH).

24.7 Example Formulation - Isotonic Energy Hydration Beverage

An isotonic energy hydration beverage illustrates how carbohydrates, electrolytes, acids, and flavor systems combine to deliver balanced hydration and sustained energy. This example highlights the rationale behind ingredient choices, sweetness targets, and osmolality design, offering a practical blueprint for building performance-focused formulations with broad consumer appeal.

Table 24.7: Benchmark Formulation: Isotonic Energy Hydration Beverage (Functional Sweetness System)

Ingredient	Function	% w/w
Purified water	Solvent	q.s. to 100
Citric acid	Acidity control	0.10
Sodium citrate	Buffer	0.25
Potassium chloride	Electrolyte	0.10
Magnesium lactate	Electrolyte	0.05
Isomaltulose	Slow-release carbohydrate	0.50
Allulose	Natural sweetness	0.30
Stevia (Reb M)	High-intensity sweetener	0.012
Natural citrus flavor	Top note	0.05
Gum Arabic	Stabilizer	0.05

Performance Metrics:

- pH: 3.8
- Osmolarity: 290 mOsm/kg (isotonic)
- Sweetness: equivalent to 5% sucrose

- Energy yield: ~1.5 kcal/100 mL
- Shelf life: 12 months (ambient, light-protected)

24.8 Formulation Insight

Practical insight emerges when sweetener selection, carbohydrate load, osmolality, and sensory goals are viewed as interconnected variables rather than isolated choices. Recognizing how each component influences hydration efficiency, flavor delivery, and stability enables formulators to refine their systems with precision. These insights guide the creation of sweetening strategies that enhance both functional performance and overall beverage experience.

- Think of sweetness as a functional nutrient, not a flavor fix.
- Match sweetness curve to the beverage's hydration tempo: quick for isotonic, slow for endurance or recovery.
- Combine sugar alcohols + rare sugars + natural sweeteners for full-palate satisfaction without metabolic load.
- Control pH and temperature during blending, they shape sweetness perception more than concentration alone.
- A successful sweetness system disappears behind the experience, leaving only clean, refreshing satisfaction.

25. Preservation Systems and Microbial Stability

Preservation is one of the most crucial, and often underestimated, dimensions of beverage formulation.
In functional electrolyte and nutrition drinks, the challenge is to achieve microbial safety, chemical stability, and flavor integrity without compromising clean-label expectations or nutritional activity.

This chapter explains how to engineer shelf stability using natural preservation systems, acidification control, and hurdle technologies that complement aseptic or cold-fill manufacturing.

25.1 The Purpose of Preservation in Functional Beverages

Preservation serves as the safeguard that protects a beverage's sensory quality, nutritional integrity, and microbial safety from production to consumption. Functional beverages, often rich in electrolytes, botanicals, or natural flavors, are especially vulnerable to degradation without a well-designed preservation strategy. Understanding the purpose of preservation provides the foundation for building drinks that remain stable, safe, and true to their intended performance throughout their shelf life.

It addresses three main degradation vectors:

1. Microbial Contamination, Yeasts, molds, bacteria, or spores.
2. Chemical Instability, Oxidation, hydrolysis, or vitamin degradation.
3. Physical Deterioration, Sedimentation, color fading, phase separation.

Formulator's Insight:
Microbial stability is the foundation of commercial shelf life. A product that is nutritionally perfect but microbially unstable will fail before it reaches consumers.

25.2 Primary Preservation Mechanisms

Preservation relies on fundamental mechanisms that disrupt microbial growth, control oxidation, and maintain chemical stability. These mechanisms, pH reduction, water-activity control, thermal treatment, and selective use of antimicrobial compounds, collectively create an environment in which spoilage organisms cannot thrive. Mastering these core principles allows formulators to engineer protection directly into the beverage matrix.

Functional drinks rely on one or more of the following preservation mechanisms:

Table 25.2: Primary Preservation Mechanisms and Functional Integration in Beverage Systems

Mechanism	Control Principle	Typical Application
pH Control (Acidification)	Inhibits microbial enzyme systems	Citric, malic, or lactic acid; pH < 4.2
Water Activity (a_n)	Reduces available moisture for growth	Carbohydrate or polyol addition
Thermal Processing	Destroys vegetative microbes	HTST or pasteurization
Chemical Preservation	Inhibits microbial reproduction	Natural or synthetic antimicrobials
Oxygen Control	Limits oxidative and microbial reactions	Nitrogen flushing, CO_2 headspace
Packaging Barrier	Prevents external contamination	Multilayer PET, aseptic seals

Best Practice:

Combine two or more mild hurdles rather than relying on one aggressive preservation step, this minimizes flavor damage and maintains nutrient potency.

25.3 Natural Preservation Systems

Natural preservation systems draw on plant extracts, organic acids, essential oils, fermentation metabolites, and mineral-based inhibitors to provide antimicrobial and antioxidant effects without synthetic additives. Their efficacy is deeply influenced by pH, solubility, flavor profile, and interactions with electrolytes or botanicals. Leveraging these systems requires balancing clean-label expectations with functional performance and sensory compatibility.

Table 25.3: Natural Preservation Systems: Sources, Mechanisms, and Clean-Label Applications

Natural Preservative	Source	Effective pH Range	Function / Mechanism	Use Level (%)	Notes
Citric Acid	Citrus fruits	2.5–4.5	pH control, chelation	0.05–0.15	Enhances others' efficacy
Ascorbic Acid	Fruits, vitamin C	3.0–5.0	Antioxidant, microbial inhibition	0.02–0.05	Sensitive to light/heat
Malic Acid	Apple, pear	3.0–5.0	Enhances acid flavor, pH control	0.05–0.10	Smooth sourness
Lactic Acid	Fermentation	3.0–4.0	Inhibits spoilage bacteria	0.05–0.10	Adds mild tang
Sorbic Acid / Potassium Sorbate	Fermentation	3.0–5.5	Mold and yeast inhibitor	0.02–0.10	Clean-label acceptable in small doses
Benzoic Acid /	Benzoin resin	2.5–4.0	Yeast & mold inhibitor	0.02–0.10	Limit in natural

Sodium Benzoate					claims regions
Rosemary Extract	Botanical	3.0–7.0	Lipid oxidation inhibitor	0.02–0.05	Adds mild herbal tone
Green Tea Polyphenols	Plant	3.0–6.0	Antioxidant, mild antimicrobial	0.01–0.03	Combine with acid system
Fermented Sugar Extracts (Nisin, Natamycin)	Microbial peptides	3.0–7.0	Bacteriostatic and antifungal	0.0005–0.01	Label as “fermented preservative”

Formulation Tip:

Combine sorbic acid (broad-spectrum) with citric + ascorbic acids (pH and antioxidant control) for a powerful yet label-friendly preservation base.

25.4 Synthetic Preservation Systems (For Comparison)

Synthetic preservatives offer predictable, robust antimicrobial activity that remains effective across diverse pH ranges, ingredient loads, and processing conditions. Their stability and reliability make them valuable benchmarks when evaluating natural alternatives. Understanding the performance characteristics of conventional preservatives helps formulators compare efficacy, determine risk tolerance, and design systems that meet both regulatory and consumer expectations.

Used in regions where natural systems are not mandated, typically for mass-market isotonic or vitamin beverages.

Table 25.4: Synthetic Preservation Systems: Efficacy, Regulatory Status, and Comparative Use Cases

Preservative	Effective Range	Function	Note
Sodium Benzoate	pH < 4.0	Yeast, mold inhibitor	Cost-effective, regulated limits
Potassium Sorbate	pH < 5.5	Yeast, mold inhibitor	Synergistic with benzoate
Calcium Disodium EDTA	Neutral	Chelator, prevents oxidation	Limited use in clean-label
Dimethyl Dicarbonate (DMDC)	Broad	Cold sterilant (microbiocidal)	Requires controlled handling

Regulatory Note:

Even when synthetic preservatives are allowed, the global trend is to replace them with acid + antioxidant + natural extract systems for "naturally preserved" positioning.

25.5 Hurdle Technology in Beverage Preservation

Hurdle technology strengthens preservation by combining multiple protective factors, such as acidity, temperature control, natural antimicrobials, and packaging barriers, into a synergistic system. When carefully layered, these hurdles reduce reliance on any single preservative and improve overall product safety and shelf stability. This integrated approach allows functional beverages to maintain quality even under variable distribution and storage conditions.

Example Hurdle Stack:

1. Low pH (3.6–4.0) → suppresses bacterial growth.
2. Antioxidant (ascorbic acid) → protects nutrients, controls oxidation.
3. Natural antimicrobial (sorbic acid / rosemary extract) → inhibits molds and yeasts.

4. Nitrogen flushing → reduces oxygen and microbial respiration.
5. UV-protected packaging → prevents photodegradation.

Result:

Long-term stability (≥12 months) without harsh chemical preservatives or pasteurization burn.

25.6 Process-Linked Preservation Strategies

Processing choices, from pasteurization and hot-fill to membrane filtration and high-pressure processing, strongly influence microbial stability and the longevity of functional beverages. Each method shapes ingredient behavior, flavor integrity, and nutrient retention differently. Aligning preservation strategy with processing conditions ensures that microbial safety is achieved without compromising sensory or functional attributes.

Table 25.6: Process-Linked Preservation Strategies and Shelf-Life Outcomes

Processing Method	Temperature	Shelf Life	Best Used For	Notes
Hot-fill	85–95 °C, 30–60 s	9–12 months	Acidic beverages	Can degrade vitamins and flavors
Aseptic Cold-fill	<25 °C, sterile filling	12–18 months	Nutrient-rich beverages	Requires high sanitation control
HTST Pasteurization	72–85 °C, 15–30 s	6–9 months	Functional waters	Good flavor retention
HPP (High-Pressure Processing)	400–600 MPa	3–6 months (refrigerated)	Premium cold-pressed drinks	Preserves natural color & aroma

UV / Pulsed Light	Non-thermal	3–6 months	Clear beverages	For small-batch or local distribution

Formulator's Insight:

For clean-label shelf-stable drinks, the most balanced route is HTST pasteurization + hurdle preservation + light-barrier packaging.

25.7 Packaging Influence on Microbial Stability

Packaging acts as an active partner in the preservation system, controlling oxygen ingress, light exposure, headspace composition, and barrier properties that directly affect microbial and oxidative stability. Material selection, sealing integrity, and fill strategy all determine how effectively the beverage is protected once it leaves production. Understanding packaging influence allows developers to extend shelf life and reduce reliance on chemical preservatives.

Table 25.7: Packaging Influence on Microbial and Chemical Stability

Packaging Material	Oxygen Barrier	Light Protection	Shelf-Life Impact
Multilayer PET	High	Medium	Standard for isotonic drinks
Amber PET / Glass	High	High	Ideal for botanical beverages
Aluminum Can	Excellent	Complete	Longest shelf life
Aseptic Carton	Very high	Moderate	Eco-friendly, neutral taste
Flexible Pouch	Variable	Low	Short distribution cycle only

Headspace Management:

- Nitrogen flushing reduces O_2 to <1 ppm, extending antioxidant and color stability.
- CO_2 sparging adds slight effervescence for flavor brightness and microbial inhibition.

25.8 Example Formulation - Naturally Preserved Functional Water

A naturally preserved functional water demonstrates how clean-label acids, mineral systems, and botanical antimicrobials can be combined with optimized pH and packaging to achieve microbial safety without synthetic preservatives. This example highlights the formulation logic, ingredient interactions, and stability considerations required to build a clear, refreshing product that meets modern consumer expectations for simplicity and naturality.

Table 25.8: Example Formulation of Naturally Preserved Functional Electrolyte Water

Ingredient	**Function**	**% w/w**
Purified water	Solvent	q.s. to 100
Citric acid	Acidulant	0.10
Sodium citrate	Buffer	0.25
Potassium sorbate	Natural preservative	0.05
Ascorbic acid	Antioxidant	0.03
Rosemary extract	Antioxidant / antimicrobial	0.02
Natural lemon flavor	Sensory	0.05
Erythritol	Sweetness	0.40
Gum Arabic	Stabilizer	0.05

Performance Metrics:

- pH: 3.8
- Total plate count: <10 CFU/mL (after 12 months, 25 °C)
- Color and aroma retention: >90%
- Preservative declaration: "Naturally preserved with plant extracts and vitamin C."

25.9 Validation and Shelf-Life Testing

Validation and shelf-life testing confirm whether a preservation system performs as intended under real-world conditions. Through microbiological challenge studies, accelerated aging, sensory monitoring, and chemical stability assays, developers can identify weak points, quantify degradation kinetics, and refine the formulation. Robust validation ensures compliance, reliability, and consumer safety across the product's lifecycle.

Every preservation system must be validated using standardized testing:

Table 25.9: Shelf-Life Validation Protocol: Microbial, Chemical, and Sensory Benchmarks

Test	Purpose	Method	Acceptance Criteria
Microbial Challenge Test	Verify preservative efficacy	Inoculate beverage with mixed flora (10^6 CFU/mL)	≥99.9% reduction after 7 days
Accelerated Stability (40 °C / 75% RH)	Predict chemical and microbial shelf life	3 months	No growth, color $\Delta E < 2$
Real-Time Shelf-Life Study	Confirm retail viability	Ambient and refrigerated	No pH drift >0.1; stable flavor profile

25.10 Formulation Insight

Effective preservation emerges from aligning ingredient choice, processing method, and packaging design into a cohesive protective system. Key insights center on recognizing how pH, ionic strength, botanical actives, and oxygen exposure interact to shape stability outcomes. Applying these learnings allows formulators to fine-tune preservation strategies that deliver safety, clarity, flavor integrity, and shelf-life performance without compromising product intent or consumer appeal.

- The best preservation strategy is layered, natural, and invisible to the consumer.
- Always design for microbial robustness + nutrient protection + sensory neutrality.
- Validate every preservative system under actual packaging and distribution conditions.
- For clean-label beverages, citric + ascorbic + natural antimicrobial remains the global gold standard.
- Preservation is not about preventing spoilage; it's about protecting value.

26. Packaging, Oxygen Control, and Shelf-Life Engineering

Packaging is not merely a container, it is an active stability system.
In modern electrolyte and nutrition beverages, packaging governs oxygen exposure, light protection, flavor preservation, and nutrient retention throughout the product's lifecycle.

Even the best formulation can degrade within weeks if oxygen or light ingress exceeds tolerance levels.
This chapter explores how to engineer package–product compatibility, optimize oxygen control, and design for target shelf life across global distribution conditions.

26.1 Functional Role of Packaging in Beverage Stability

Packaging plays an active, functional role in beverage stability by protecting the product from oxygen, light, temperature fluctuation, and mechanical stress. Beyond its structural purpose, packaging determines how effectively a beverage maintains its flavor integrity, nutrient potency, color stability, and microbial safety throughout its lifecycle. Treating packaging as a stability component—not an afterthought—ensures the formulation performs consistently from production to consumption.

Packaging performs three simultaneous roles:

1. Barrier Function, Prevents oxygen, moisture, and UV ingress.
2. Containment Function, Maintains carbonation, aseptic conditions, and structure.
3. Communication Function, Conveys freshness, health, and sustainability to the consumer.

Formulator's Insight:

Packaging should be designed in parallel with formulation, not as a final marketing step. Beverage chemistry dictates what the package must protect against.

26.2 Key Degradation Pathways Controlled by Packaging

Oxygen ingress, UV-induced degradation, aroma loss, nutrient oxidation, and microbial exposure are among the primary degradation pathways influenced directly by packaging. The permeability, reflectivity, and barrier strength of the chosen material dictate how well these pathways are controlled. Proper alignment between packaging characteristics and formulation vulnerabilities is crucial to preventing instability, discoloration, and off-flavor development over time.

Table 26.2: Key Degradation Pathways in Functional Beverages and Packaging-Based Mitigation Strategies

Degradation Type	Cause	Packaging Solution
Oxidation	Oxygen ingress through headspace or wall	Oxygen scavengers, nitrogen flushing, barrier layers
Photodegradation	UV/visible light exposure	Amber or opaque materials
Aroma Loss	Permeation or sorption	Multilayer PET, glass, or lined aluminum
Nutrient Degradation	Vitamin C, polyphenol oxidation	O_2 barrier + low light + antioxidant system
Carbonation Loss	CO_2 permeation	Glass, aluminum, CO_2 barrier PET
Microbial Growth	Recontamination	Hermetic sealing, aseptic filling

Shelf-Life Equation:

Shelf Life = f (Oxygen Transmission + Light Exposure + Headspace O2 + Process Hygiene)

26.3 Oxygen Management Strategies

Oxygen management is critical for preserving sensitive components such as natural flavors, vitamins, oils, and pigments. Strategies include selecting low-permeability materials, incorporating oxygen scavengers, minimizing headspace oxygen during filling, and optimizing sealing integrity. Effective oxygen control extends shelf life by slowing oxidative reactions and maintaining the beverage's intended sensory and nutritional profile.

Oxygen is the primary enemy of beverage stability, it drives vitamin loss, color fading, and flavor off-notes.

Sources of Oxygen:

- Dissolved in water or ingredients.
- Trapped in headspace during filling.
- Permeating through packaging walls.

Control Strategies:

Table 26.3: Oxygen Management Strategies for Extended Shelf-Life Performance

Control Step	Method	Effectiveness
Deaeration	Vacuum or membrane degassing of water	Removes 95–99% dissolved O_2
Nitrogen Flushing	Displaces O_2 in headspace	Reduces headspace $O_2 < 0.5\%$
Oxygen Scavengers	Reactive sachets or polymer layers	Continuous O_2 capture
Low-Permeability Materials	Multilayer PET, EVOH barrier	Prevents O_2 ingress
Antioxidant Coupling	Ascorbic acid, rosemary extract	Neutralizes residual O_2

Best Practice:

Measure dissolved oxygen (DO) immediately before capping, ideal DO levels for long shelf life are <0.5 mg/L.

26.4 Light Protection and Color Retention

Light exposure—particularly UV and high-energy visible light—accelerates degradation of natural colors, aromas, and nutrients. Packaging solutions such as tinted bottles, multilayer materials, and UV-blocking films help preserve color intensity and sensory quality. Matching light protection to the specific sensitivity of the formulation ensures long-term stability and visual appeal.

Light exposure (particularly UV < 400 nm) triggers oxidation and pigment breakdown.
For natural colorants (anthocyanins, curcumin, carotenoids), even 72 hours of exposure to 1000 lux can reduce color intensity by 30%.

Table 26.4: Light Transmission Characteristics of Common Packaging Materials

Packaging Type	Light Transmission (400–700 nm)	Recommended Use
Clear PET	80–90%	Short shelf life, refrigerated
Amber PET	<10%	Standard functional waters
Opaque PET (white/black)	<1%	Isotonic or protein beverages
Aluminum Can	0%	Maximum protection
Dark Glass	<5%	Premium antioxidant beverages

Formulator's Insight:

Match color sensitivity to package opacity: anthocyanin and chlorophyll systems require amber or opaque packaging for shelf stability beyond six months.

26.5 Packaging Material Comparison

Different packaging materials—PET, HDPE, glass, aluminum, and multilayer composites—offer varying levels of oxygen and light protection, recyclability, cost efficiency, and consumer experience. PET provides good clarity and cost-effective performance, glass offers excellent barrier properties, aluminum excels in light protection, and multilayer structures deliver tailored barrier profiles. Selecting the appropriate material requires balancing functional needs with sustainability and brand positioning.

Table 26.5: Packaging Material Comparison: Barrier Performance, Sustainability, and Application Fit

Material	O_2 Barrier	Light Barrier	Weight	Recyclability	Key Advantage	Limitation
Glass	Excellent	Excellent (amber)	High	High	Premium, inert	Fragile, heavy
Aluminum Can	Excellent	Complete	Medium	High	Best barrier + portability	Not transparent
Multilayer PET (EVOH)	Very good	Moderate	Low	Moderate	Lightweight, transparent	Barrier degrades under humidity
Aseptic Carton	Very good	High	Low	Medium	Eco perception, light block	Limited heat tolerance

Monolayer PET	Moderate	Poor	Low	High	Economical	High O_2 ingress
Flexible Pouch (foil)	Variable	High	Very low	Medium	Lightweight, resealable	Short shelf life, permeability over time

26.6 Headspace and Gas Control

Headspace composition strongly influences oxidation rate, carbonation behavior, and aromatic stability. Managing dissolved gases, removing entrapped air before filling, and using nitrogen or CO_2 headspace flushing can significantly improve shelf life. Precise gas control is especially important for sensitive beverages containing antioxidants, natural flavors, or high electrolyte loads.

Proper headspace management stabilizes both flavor and structure.

Table 26.6: Headspace Gas Control Methods and Functional Outcomes

Method	Gas Used	Target Level	Benefit
Nitrogen Flushing	N_2	99.9% purity	Displaces oxygen, maintains pressure
Carbon Dioxide Sparging	CO_2	1–2 vol%	Adds light effervescence and microbial protection
Vacuum Capping	-	$<1\%$ O_2 headspace	Minimizes oxidation for non-carbonated drinks

Practical Tip:

For still functional waters, maintain headspace ≤ 3% of total volume; excess headspace accelerates oxygen ingress.

26.7 Shelf-Life Modeling and Accelerated Testing

Shelf-life modeling uses controlled temperature, light, and oxygen conditions to predict long-term product performance, allowing early identification of failure modes. Accelerated testing speeds up degradation pathways while providing actionable data on nutrient retention, sensory stability, and physical integrity. These tools guide packaging decisions and help optimize preservation strategies before full-scale launch.

Predictive shelf-life modeling combines oxygen ingress, temperature, and light exposure data to estimate real-world product stability.

Table 26.7: Shelf-Life Modeling and Accelerated Testing Protocols

Test	Purpose	Condition	Duration	Indicator
Accelerated Aging	Predict degradation rate	40 °C, 75% RH	3 months	Nutrient & color loss
Thermal Cycling	Simulate transport stress	4–40 °C cycles	30 days	Haze, pH shift
Light Exposure Test	Photostability	1000 lux, 25 °C	14 days	Color ΔE < 2
O_2 Transmission Test	Barrier performance	ASTM D3985	Continuous	OTR < 0.1 cc/m²/day

Shelf-Life Estimation Rule:

For every 10 °C increase, reaction rates roughly double (Arrhenius principle). Thus, 3 months at 40 °C ≈ 12 months at 25 °C for prediction purposes.

26.8 Sustainability and Barrier Innovations

Sustainable packaging innovation focuses on reducing environmental impact while maintaining essential barrier performance. Recycled-content PET, bio-based polymers, lightweight aluminum, and advanced barrier coatings offer pathways to improved sustainability. New technologies enable circular design without compromising oxygen control, light protection, or structural integrity, supporting both ecological goals and product stability.

Modern packaging must balance technical performance with environmental responsibility.

Table 26.8: Sustainable Packaging Innovations with Functional Performance Trade-Offs

Innovation	Description	Functional Advantage
Bio-PET (plant-based)	Derived from renewable sources	Same performance, reduced CO_2 footprint
Recycled PET (rPET)	Circular material use	Moderate barrier, eco appeal
Active Barrier Additives	Oxygen scavenging polymers (e.g., MXD6, polyamide blends)	Continuous protection
Compostable Films	PLA blends with coatings	Short shelf-life beverages
Smart Packaging	Oxygen or freshness indicators	Consumer transparency
UV-Blocking Coatings	Layered barrier coatings	Color and vitamin protection

Sustainability Tip:

Balance "eco" with performance, a sustainable pack that allows oxidation is not sustainable if it causes early spoilage.

26.9 Example Application – Oxygen-Controlled Electrolyte Beverage

An oxygen-controlled electrolyte beverage illustrates how low-permeability materials, nitrogen flushing, and tight sealing parameters work together to protect sensitive flavors and electrolytes from oxidation. This example demonstrates the importance of aligning packaging with formulation needs, resulting in a product that maintains clarity, freshness, and functional integrity across its full shelf life.

Table 26.9: Example Formulation: Oxygen-Controlled Electrolyte Beverage with Packaging Integration

Ingredient	Function	% w/w
Water (deaerated)	Solvent	q.s. to 100
Citric acid	pH adjustment	0.10
Sodium citrate	Buffer	0.25
Magnesium chloride	Electrolyte	0.05
Ascorbic acid	Antioxidant	0.03
Natural lemon oil (encapsulated)	Flavor	0.05
Erythritol	Sweetener	0.40
Potassium sorbate	Preservative	0.05

Packaging System:

- Amber multilayer PET with EVOH oxygen barrier layer.
- Nitrogen flushing to <0.5 mg/L dissolved O_2.
- Foil induction seal + tamper-evident cap.

Performance Metrics:

- Oxygen ingress: <0.05 cc/m²/day
- Vitamin C retention: 88% after 12 months
- Color ΔE: <1.5 (stable visual)
- Shelf life: 12–15 months ambient

26.10 Formulation Insight

Insight arises when packaging is integrated early into the formulation process, ensuring that barrier properties, fill conditions, and stability goals are aligned. Recognizing how packaging interacts with oxygen-sensitive nutrients, natural colors, and volatile aromas allows developers to tailor stability strategies more effectively. This holistic approach produces beverages that remain consistent, high-performing, and shelf-stable under real-world conditions.

- Packaging is a formulation extension; every nutrient and color molecule "feels" its barrier environment.
- Prioritize oxygen management first, light protection second, and sustainability third, in that order for shelf-life control.
- The smallest detail (e.g., headspace gas purity or cap permeability) can shorten stability by months.
- For export beverages, always perform full transport simulation, vibration, temperature, and light cycles can alter barrier performance.
- Sustainable doesn't mean fragile, it means durable protection with responsible materials.

27. Processing Technologies and Manufacturing Integration

Processing defines the bridge between formulation and reality.
Even the most stable, well-balanced formula can fail in scale-up if thermal, shear, or aeration conditions are not aligned with ingredient sensitivities.

This chapter presents the engineering foundations and practical optimization strategies for processing functional electrolyte and nutrition beverages, from thermal pasteurization and homogenization to modern non-thermal stabilization systems such as HPP and UV-C treatment.

27.1 The Role of Processing in Functional Beverage Design

Processing shapes the physical, chemical, and sensory outcomes of a functional beverage just as profoundly as the ingredients themselves. Heat, pressure, shear, and oxygen exposure all influence stability, clarity, nutrient retention, and flavor. Understanding the role of processing provides the foundation for designing beverages that remain consistent and high-performing from batch to bottle.

Beverage processing must simultaneously achieve:

1. Microbial Safety – eliminate pathogens and spoilage organisms.
2. Physical Stability – prevent sedimentation, creaming, or phase separation.
3. Nutrient Retention – preserve vitamins, flavors, and bioactives.
4. Sensory Integrity – maintain desired taste, color, and aroma profile.

Processing is therefore a balance of thermal energy, mechanical energy, and time, orchestrated to meet regulatory and sensory targets.

27.2 Key Processing Stages in Beverage Manufacturing

Core processing stages, mixing, heating, homogenization, filtration, filling, and cooling, each impose distinct stresses on ingredients and structural systems. How these steps are sequenced and controlled determines the final product's stability, safety, and sensory balance. Mapping these stages enables developers to anticipate risks and engineer a more robust manufacturing workflow.

Table 27.2: Key Processing Stages in Functional Beverage Manufacturing and Critical Control Parameters

Step	Purpose	Key Variables	Common Challenges
Ingredient Preparation	Dissolution, dispersion	Mixing speed, hydration time	Foaming, incomplete solubility
Filtration / Clarification	Remove solids, ensure clarity	Pore size, pressure	Loss of color or aroma
Deaeration	Remove dissolved oxygen	Vacuum level, temperature	Oxygen ingress post-process
Homogenization	Particle size reduction	Pressure (30–150 bar), passes	Over-shearing delicate emulsions
Thermal Treatment	Microbial inactivation	Temperature, time	Nutrient degradation
Filling & Sealing	Maintain sterility	Headspace O_2, temperature	Seal integrity, recontamination

Formulator's Insight:

Design your formula for the process, not after it.
Thermal tolerance, pH, and viscosity all determine whether a beverage can survive hot-fill, cold-fill, or HPP conditions.

27.3 Thermal Processing Options

Thermal treatments such as pasteurization, hot-fill, and UHT processing establish microbial safety while also influencing flavor, nutrient stability, and color. Each method carries unique advantages and trade-offs in terms of heat load and product impact. Selecting the right thermal option ensures safety without compromising functional or sensory quality.

Thermal methods remain the backbone of beverage safety, simple, proven, and widely regulated.

Table 27.3: Thermal Processing Options for Hydration and Nutrition Beverages

Method	Temperature (°C)	Time	Typical Application	Advantages	Limitations
Hot-Fill	85–95	30–60 s	Acidic functional drinks	Simple, cost-effective	Flavor and vitamin loss
HTST (High-Temp Short-Time)	72–85	15–30 s	Clear hydration or isotonic	Balanced stability and taste	Needs pasteurizer unit
UHT (Ultra-High Temp)	135–150	1–3 s	Protein or dairy-based drinks	Long shelf life	Flavor "cooked" notes
Tunnel Pasteurization	60–75	15–25 min	Canned or bottled drinks	Post-fill sterilization	Energy intensive
Batch Pasteurization	63–65	30 min	Small batches	Low cost	Low throughput, less control

Optimization Rule:

Every 10 °C reduction in temperature roughly doubles vitamin retention, aim for the shortest, mildest heat treatment compatible with microbial safety.

27.4 Non-Thermal Processing Technologies

Emerging non-thermal techniques, including high-pressure processing, pulsed electric fields, membrane filtration, and UV treatment, offer alternatives that preserve freshness and nutrient integrity while delivering strong microbial reduction. These technologies minimize heat stress and broaden formulation possibilities for sensitive ingredients and clean-label systems.

To meet the "fresh" and "cold-pressed" trend, non-thermal preservation methods are increasingly applied to functional drinks.

Table 27.4: Non-Thermal Processing Technologies for Premium Functional Beverages

Technology	Mechanism	Advantages	Limitations	Best Use
HPP (High Pressure Processing)	400–600 MPa destroys microbial membranes	Retains flavor, color, nutrients	High equipment cost, cold chain required	Premium cold-pressed or adaptogenic drinks
Pulsed Electric Field (PEF)	Short electric bursts disrupt cells	Minimal heat, good nutrient retention	Not yet global scale	Clear juices, waters
UV-C Treatment	Germicidal light destroys DNA	Non-thermal, continuous flow	Limited penetration in turbid liquids	Clear beverages only
Membrane Filtration (Micro / Nano)	Physical removal of microbes	Low energy, high clarity	Potential flavor loss	Electrolyte and protein waters

Cold Plasma	Ionized gas sterilization	Surface decontamination	Experimental	Bottle sterilization

Formulator's Insight:

Non-thermal systems excel for high-value, delicate beverages where freshness and nutrient retention outweigh long ambient shelf life.

27.5 Homogenization and Emulsion Stability

Homogenization plays a central role in creating fine-droplet emulsions, improving suspension, and ensuring long-term stability of flavor oils and color dispersions. Pressure settings, shear intensity, and equipment design all influence droplet size distribution and physical resilience. Mastery of homogenization helps create smooth, uniform, and shelf-stable beverages.

Homogenization reduces particle size and ensures uniform distribution of emulsified flavors, vitamins, or lipids.

Table 27.5: Homogenization Parameters for Flavor, Protein, and Stabilizer Systems in Functional Beverages

Application	Pressure (bar)	Passes	Result
Flavor emulsions	60–120	2	Stable dispersion, <1 μm droplets
Protein drinks	100–150	1–2	Smooth texture, no creaming
Gum-stabilized waters	30–60	1	Uniform mouthfeel

Critical Factors:

- Use low-shear mixing before homogenization to avoid aeration.
- Maintain temperature <50 °C for flavor-loaded emulsions.
- Add antioxidants prior to homogenization to prevent oxidative initiation.

27.6 Deaeration and Oxygen Removal

Oxygen removal is essential for preventing oxidation-driven flavor loss, color degradation, and vitamin breakdown. Deaeration systems, through vacuum, sparging, or membrane technologies, reduce dissolved oxygen prior to filling, improving overall shelf life. Effective control of oxygen enhances stability from both a sensory and chemical perspective.

Dissolved oxygen not only oxidizes sensitive ingredients (vitamin C, carotenoids) but also reduces microbial preservative efficiency.

Table 27.6: Deaeration Methods and Oxygen Removal Efficiency for Hydration Beverage Systems

Method	Mechanism	Efficiency	Integration
Vacuum Deaeration	Negative pressure removes O_2	95–99%	Standard for large plants
Membrane Deaeration	Hydrophobic membrane diffuses O_2 out	>99%	Compact and efficient
CO_2 Sparging	Displacement of O_2 by CO_2 bubbles	90–95%	Adds light effervescence
N_2 Injection	Gas stripping	90–98%	Enhances oxygen barrier and headspace control

Rule of Thumb:

Dissolved oxygen after processing should be <0.5 mg/L for vitamin-rich formulations.

27.7 Integration of Preservation and Processing ("Hurdle Engineering")

Processing and preservation work best when integrated into a layered system, where acidity, heat treatment, oxygen management, and packaging barriers reinforce each other. This "hurdle engineering" approach minimizes reliance on any single method and creates a more resilient, predictable stability profile. Strategic integration ensures safety, quality, and longevity across diverse formulations.

Processing and preservation must be viewed as a unified stability ecosystem rather than isolated steps.

Example Hurdle Stack:

1. Deaeration – 99% O_2 removal.
2. Acidification – pH 3.8 ± 0.1.
3. HTST Pasteurization – 72 °C for 20 s.
4. Antioxidant addition – Ascorbic acid + rosemary extract.
5. Nitrogen flushing + multilayer PET packaging.

Result:

12-month ambient stability with >85% vitamin retention and <1 NTU clarity deviation.

27.8 Process Flow Example – Functional Hydration Beverage

A functional hydration beverage process flow illustrates how ingredient dissolution, thermal treatment, homogenization, deaeration, and controlled filling combine to produce a stable and refreshing product. This example highlights the logic behind each step and demonstrates how process design reinforces formulation goals.

1. Water Preparation
   - Filtration (5 µm) → Deaeration (vacuum 0.09 MPa).
2. Dry Ingredient Blending
   - Dissolve salts and acids at 30–40 °C under moderate agitation.
3. Sweetener Integration
   - Add carbohydrates and natural sweeteners; maintain pH 3.6–3.8.
4. Flavor and Color Addition
   - Add encapsulated or emulsified flavors under low shear.
5. Homogenization
   - 2 passes at 80 bar for uniform dispersion.
6. Pasteurization (HTST)
   - 80 °C for 25 s; cool immediately to 25 °C.
7. Filling & Nitrogen Flushing
   - Maintain headspace $O_2 < 0.5\%$.
8. Packaging
   - Amber PET (EVOH barrier) or aluminum cans.
9. Storage Validation
   - Accelerated at 40 °C / 75% RH for 3 months.

Outcome:

Stable isotonic beverage with pH drift <0.05, color $\Delta E <1.2$, and zero microbial growth at 12 months.

27.9 Common Processing Pitfalls and Solutions

Processing challenges, such as foaming, flavor degradation, destabilized emulsions, vitamin loss, and precipitation, often arise from small mismatches between ingredient behavior and process conditions. Identifying these pitfalls and applying targeted solutions helps safeguard consistency, improve efficiency, and prevent costly production failures.

Table 27.9: Common Processing Pitfalls in Functional Beverage Manufacturing and Corrective Actions

Issue	Likely Cause	Corrective Action
Flavor degradation	Excessive thermal load	Switch to HTST or HPP
Vitamin loss	High residence time or O_2	Add antioxidants post-pasteurization
Foaming / air entrapment	High agitation or surfactant imbalance	Vacuum deaerate, control mixing speed
Phase separation	Poor homogenization or gum ratio	Adjust pressure or stabilizer blend
Sedimentation	Particle aggregation	Use low-acyl gellan or microfluidization

27.10 Quality Control Checkpoints During Processing

Quality control checkpoints monitor critical parameters such as pH, viscosity, droplet size, dissolved oxygen, temperature, and microbial load throughout manufacturing. These checkpoints ensure that each stage meets specification before moving forward, reducing variability and maintaining product integrity at scale.

Table 27.10: Critical Quality Control Checkpoints During Beverage Processing

Stage	Critical Parameter	Monitoring Method	Target Value
Water feed	Conductivity / DO	Inline sensor	<0.5 mg/L O_2
Mixing	pH, viscosity	pH meter, viscometer	pH 3.6–4.0
Pasteurization	Temperature–time	Data logger	80 ± 1 °C / 25 s

Filling	Headspace O_2	Inline analyzer	<0.5%
Final Product	Microbial load	Plate count	<10 CFU/mL

27.11 Formulation Insight

Insight emerges when processing conditions are viewed not as constraints but as active design variables that shape stability, flavor, and nutritional performance. Recognizing how heat, shear, oxygen, and mixing dynamics interact with formulation components allows developers to refine both the beverage and the process with greater precision.

- Process defines shelf life. A great formula cannot compensate for oxygen or heat abuse.
- Treat processing variables as formulation ingredients, they modify chemistry just as pH or flavor does.
- Always conduct pilot-scale validation before full-scale production to capture real shear, oxygen, and dwell-time effects.
- When targeting premium or "fresh" positioning, explore HPP or PEF, but verify cost and distribution feasibility.
- The best process is one that delivers safety invisibly, preserving freshness without consumers ever knowing it happened.

28. Regulatory Compliance, Labeling, and Nutrient Claims

Formulating a scientifically sound beverage is only half of success; the other half is regulatory and labeling compliance.

No matter how innovative a formula is, it cannot reach consumers without meeting national and international standards for safety, composition, and claim substantiation.

This chapter provides a structured roadmap for ensuring regulatory alignment, clean-label integrity, and transparent communication in functional electrolyte and nutrition beverages across the U.S., EU, and emerging markets.

28.1 The Purpose of Regulatory Compliance

Regulatory compliance ensures that functional beverages are safe, accurately represented, and legally marketable across their intended regions. Compliance protects consumers, supports brand integrity, and defines the boundaries within which innovation can occur. Understanding its purpose helps developers design products that meet both regulatory expectations and commercial goals.

Regulatory systems exist to ensure that:

1. Consumers receive truthful, non-misleading information.
2. Ingredients and additives meet safety and purity standards.
3. Health claims are scientifically substantiated.

For beverage developers, compliance is both a quality management tool and a brand protection strategy, it prevents recalls, protects credibility, and builds consumer trust.

28.2 Global Regulatory Framework Overview

Global beverage regulations vary widely, but all share common foundations around safety, labeling accuracy, and permissible ingredient use. Mapping these frameworks: U.S., EU, Asia-Pacific, Middle East, and Latin America, provides clarity on how formulations must adapt for different markets. Awareness of these structures allows teams to plan for international expansion without reformulation surprises.

Table 28.2: Global Regulatory Framework Overview for Hydration and Nutrition Beverages

Region	Primary Authority	Core Regulation	Key Compliance Focus
United States	FDA (Food and Drug Administration)	21 CFR, DSHEA (Dietary Supplement Health & Education Act)	GRAS status, structure/function claims
European Union	EFSA (European Food Safety Authority)	Regulation (EC) No 1924/2006	Nutrition & health claims approval list
United Kingdom	FSA (Food Standards Agency)	UK Nutrition & Health Claims Register	Post-Brexit parallel to EFSA
GCC & MENA	SFDA, ESMA, GSO	Gulf Standard GSO 9/2013	Ingredient approval, Arabic/English labeling
Canada	Health Canada, CFIA	Food and Drugs Act	Natural health product vs. beverage distinction
Asia-Pacific	Varies (MOH, FSSAI, etc.)	National food safety standards	Fortification limits, additive lists

Formulator's Insight:

Always classify your product correctly first, as a food, functional beverage, or dietary supplement, since labeling, additive permissions, and claims differ significantly.

28.3 Ingredient Regulatory Status and GRAS Considerations

Ingredient approval status determines whether a component can be used freely, used with restrictions, or requires additional safety documentation. GRAS determinations, positive lists, purity standards, and regional differences all shape what is allowable in a functional beverage. Understanding these classifications ensures formulations remain compliant from concept through commercialization.

In the U.S., an ingredient must be GRAS (Generally Recognized As Safe) or approved as a food additive.

In the EU, it must appear in the Union List of Approved Additives.

Table 28.3: Ingredient Regulatory Status and GRAS/Approval Pathways

Ingredient Category	Typical Regulatory Pathway	Notes
Electrolytes (Na, K, Mg, Ca salts)	GRAS / approved additives	Must use food-grade purity
Natural Flavors & Colors	FEMA-GRAS or EU Annex II	"Natural flavor" only if derived from plant/fruit
Sweeteners (stevia, erythritol, monk fruit)	GRAS (USA) / E960–E968 (EU)	Label as "sweetener from [source]"
Botanical Extracts (adaptogens, herbs)	Dietary supplement category	Must have safety dossier or traditional use record
Preservatives (sorbate, benzoate)	Approved additive (E200–E219)	Label by name or E-number

Vitamins & Minerals	Fortification regulation	Must not exceed upper safe limits

Practical Tip:

Use FDA's GRAS notices and EFSA's approved ingredient database early in formulation planning to prevent last-minute reformulation.

28.4 Nutrient and Functional Claims – Global Perspective

Claims, such as "electrolyte replenishment," "supports hydration," or "vitamin C contributes to normal immune function", are tightly regulated and differ across global markets. Requirements around evidence, wording, and nutrient thresholds determine what can legally be stated on a label. Knowing these rules ensures claims are both compelling and defensible.

Claims communicate function, but every region strictly defines what you can (and cannot) say.

Table 28.4: Nutrient and Functional Claims: Global Regulatory Landscape and Implementation Guidance

Claim Type	Example	Regulation / Requirement	Notes
Nutrient Content Claim	"High in Vitamin C"	≥20% DV (U.S.) or source threshold (EU)	Must list %DV on label
Comparative Claim	"50% less sugar than standard sports drink"	Requires identified reference	Must be statistically valid
Structure/Function Claim	"Supports hydration and muscle recovery"	U.S.: DSHEA notification to FDA	Must not imply disease prevention

Health Claim	"Calcium supports bone health"	Pre-approved by EFSA/FDA	Must use authorized wording
Prohibited Claim	"Cures fatigue" or "prevents dehydration illness"	Forbidden	Considered medical claim

Formulator's Insight:

When in doubt, rephrase functional claims into structure–function language: Instead of "prevents dehydration," say "supports optimal hydration balance."

28.5 Clean-Label and Natural Positioning Regulations

Terms like "natural," "clean-label," "no preservatives," and "minimally processed" carry varied and often strict regulatory interpretations. Understanding how each region defines and enforces these concepts allows developers to align marketing language with legal boundaries. Strong positioning begins with a clear grasp of how these terms are regulated globally.

"Natural," "organic," and "additive-free" are not interchangeable, each has unique definitions (or lack thereof).

Table 28.5: Clean-Label and Natural Positioning: Regulatory Definitions and Labeling Boundaries

Label Term	Regulatory Meaning	Key Condition
Natural Flavor	Derived from plant/fruit without chemical modification	Must comply with 21 CFR 101.22
Natural Color	Extracted from edible sources, no synthetic process	Label as "color from [source]"
Organic	Certified per USDA or EU Organic	95% or more certified ingredients

No Preservatives	Only if shelf life achieved without preservatives	Must validate via microbial challenge
Non-GMO	Verified per third-party audit	Traceability required
Plant-Based	Derived from botanical origin	Must not contain animal derivatives

Labeling Tip:

Avoid overpromising, consumers trust "minimally processed" more than "chemical-free," which regulators view as misleading.

28.6 Mandatory Labeling Elements

Labels must provide essential information, ingredient statements, allergens, nutrition panels, net contents, manufacturing details, and storage instructions, presented in specific formats. These requirements ensure transparency and consumer safety. Accurate labeling serves as a critical bridge between formulation intent and regulatory compliance.

All packaged beverages must include specific elements in the local language(s).

Standard Label Layout (Front / Back Panels):

Table 28.6: Mandatory Labeling Elements for RTD Functional Beverages

Element	Description	Notes
Product Name	Descriptive, not misleading	e.g., "Electrolyte Hydration Drink with Natural Citrus Flavor"
Net Content	Volume in mL or L	Front panel, ≥3 mm font
Ingredient List	Descending order by weight	Must declare allergens and additives
Nutrition Facts	Standardized format	kcal, macronutrients, %DV

Allergen Declaration	Must highlight allergens (soy, milk, nuts)	Bold font or separate statement
Storage Instructions	e.g., "Keep refrigerated after opening"	Required if temperature-sensitive
Manufacturer Information	Name, address, country of origin	Mandatory
Lot Number & Expiry Date	Traceability requirement	"Best before" or "Use by" date
Claims & Certifications	e.g., "Non-GMO", "Organic"	Must be verifiable
Barcode & Recycling Info	Logistics and sustainability	Optional but standard

Formulator's Insight:

Label is a legal document. Ensure version control: each reformulation or process change may require label update.

28.7 Fortification and Upper Intake Limits

Nutrient fortification must remain within established daily intake limits and adhere to permitted forms, potencies, and combinations. Excessive fortification can trigger safety concerns or restrict market access. Understanding these limits allows developers to create functional beverages that deliver meaningful benefits while remaining compliant and safe.

Functional beverages often include vitamins, minerals, or amino acids, but levels must respect regulatory tolerances.

Table 28.7: Vitamin and Mineral Fortification: Per-Serving Upper Limits and Safety Thresholds

Nutrient	U.S. RDI (%DV per 100 mL typical limit)	EFSA Upper Limit (per 250 mL)	Note
Vitamin C	≤100% DV (90 mg)	250 mg	Heat-sensitive; test retention
Vitamin B6	≤25% DV (0.5 mg)	2 mg	Excess may require warning
Magnesium	≤20% DV (80 mg)	250 mg	Check salt solubility
Zinc	≤10% DV (1 mg)	5 mg	Metallic taste at higher levels
Caffeine	≤40 mg/serving (soft drinks)	200 mg/day limit	Must declare: "Contains caffeine"
Collagen Peptides	Not standardized	-	Classify as protein source

Tip: Declare "% Daily Value per serving" clearly, regulators evaluate both per-100 mL and per-portion basis depending on jurisdiction.

28.8 International Label Adaptation

Launching a product globally requires adapting labels to different languages, unit systems, nutrition formats, allowable claims, and regulatory constraints. Effective adaptation maintains brand coherence while ensuring compliance across diverse markets. This step is essential for smooth international expansion.

To export globally, labeling and formulations must adapt to regional norms.

Table 28.8: International Label Adaptation Guidelines by Key Market

Market	Language Requirement	Unique Regulation	Packaging Metric

EU / UK	Local language + nutrition in kJ	Requires allergen bolding	250–500 mL
GCC / Middle East	Arabic + English	Date format: DD/MM/YYYY	250–350 mL typical
Asia (China, Japan)	National standard characters	Requires ingredient translation	300–500 mL
U.S. / Canada	English (French for Canada)	%DV mandatory	Fluid ounces + mL

Regulatory Strategy:

Develop a master formulation dossier that includes ingredient specifications, certificates, and local equivalents (e.g., "citric acid E330") to accelerate regional registration.

28.9 Documentation and Traceability

Documentation, specifications, certificates of analysis, hazard assessments, allergen controls, and manufacturing records, forms the backbone of regulatory compliance. Traceability systems ensure ingredients can be tracked from supplier to finished product, supporting audits and protecting consumer safety. Strong documentation prevents delays and regulatory setbacks.

Modern food regulations emphasize full traceability from ingredient sourcing to consumer.

Minimum Documentation Set:

- Certificate of Analysis (CoA) for each raw material.
- GRAS or additive approval documentation.
- Process flow diagram + HACCP plan.
- Batch production records.
- Stability and shelf-life validation reports.

- Label compliance verification (internal audit).

Practical Tip:

Maintain a digital specification system (PLM) to synchronize formulation, labeling, and regulatory compliance in real time across production sites.

28.10 Example: Label Compliance Snapshot

A label compliance snapshot illustrates how ingredient lists, nutrient panels, claims, and formatting requirements come together in a finished product. This example highlights the key decisions and checks needed to ensure a functional beverage label is accurate, compliant, and ready for market.

Front Panel (U.S. Market):

FM Healthy Beverages of America
CITRUS ELECTROLYTE WATER
Naturally flavored hydration drink with magnesium and potassium
330 mL (11.2 fl oz)

Supports hydration & refreshment

Back Panel:

Ingredients: Purified Water, Isomaltulose, Erythritol, Citric Acid, Magnesium Chloride, Potassium Citrate, Natural Citrus Flavors, Ascorbic Acid (Vitamin C), Gum Arabic, Potassium Sorbate.

Nutrition Facts (per 330 mL):
Energy: 12 kcal | Carbohydrate: 3.5 g | Sugars: 0 g | Sodium: 45 mg | Potassium: 80 mg | Magnesium: 20 mg | Vitamin C: 30 mg (33% DV)

Store in a cool, dry place. Best before: see bottle neck.

Claim: "Supports hydration and electrolyte balance."
(Structure/function claim – compliant with DSHEA 21 CFR 101.93)

28.11 Formulation Insights

Insight emerges when regulatory limitations are integrated into the early stages of formulation rather than treated as a final checkpoint. Understanding how claims, ingredient status, fortification limits, and labeling rules interact enables developers to design beverages that are innovative, functional, and fully compliant across their intended markets.

- Compliance should begin at formulation, not after packaging.
- Every additive, sweetener, and color must have a legal justification and purity record.
- Use scientifically supportable language; never promise disease prevention or cure.
- Regulatory approval is not an obstacle; it is your license for market access.
- Build your brand's reputation on transparency and precision, not exaggeration.

29. Quality Control, Analytical Testing, and Shelf-Life Validation

In beverage science, quality is not tested into the product, it is built into it. However, robust quality control (QC) and analytical testing systems ensure that every batch meets the same sensory, nutritional, and safety profile envisioned during formulation.

This chapter defines the methods, standards, and validation frameworks that guarantee batch-to-batch consistency, product integrity, and predictable shelf life for electrolyte and functional nutrition beverages.

29.1 Objectives of Quality Control in Functional Beverages

Quality control ensures that every batch of a functional beverage meets defined standards for safety, stability, potency, and sensory performance. Its objectives extend beyond compliance, providing the structure needed to maintain consistency across variable ingredients, complex formulations, and demanding processing environments. Clear QC goals form the backbone of reliable, scalable beverage production.

QC systems in beverage production serve five primary goals:

1. Safety: Verify absence of pathogens and chemical contaminants.
2. Identity: Confirm that each ingredient and active matches specification.
3. Consistency: Maintain uniform color, flavor, and physicochemical profile.
4. Stability: Ensure product retains potency, taste, and clarity through shelf life.
5. Compliance: Meet legal and labeling requirements in every market.

Formulator's Insight:

A QC system is not only a compliance tool, it's an early-warning mechanism for formulation drift, raw material variability, and process stress.

29.2 Core Analytical Parameters and Methods

Analytical testing anchors the QC process, measuring parameters such as pH, Brix, osmolality, viscosity, mineral concentration, vitamin potency, and droplet size distribution. These data points reveal how well the formulation withstands processing and guide corrective adjustments. Mastering these methods ensures precise control over both functional and sensory outcomes.

Table 29.2: Core Analytical Parameters and Methods for Beverage Quality Assurance

Parameter	Purpose	Common Method	Target / Specification
pH	Acid balance, microbial control	pH meter (calibrated daily)	3.5–4.0 typical
Brix (°Bx)	Soluble solids / sweetness check	Refractometer	±0.2°Bx of target
Turbidity / Clarity	Visual and optical quality	Nephelometer (NTU)	<1.5 NTU (clear beverages)
Color (ΔE)	Visual consistency	CIELAB spectrophotometer	ΔE < 2 vs. standard
Conductivity	Electrolyte strength	Conductivity meter	±5% of specification
Viscosity	Mouthfeel / stability	Brookfield viscometer	±10% of design
Dissolved Oxygen (DO)	Oxidation control	DO probe	<0.5 mg/L

Nutrient Assay (Vitamin C, Mg, etc.)	Potency verification	HPLC / ICP-OES	≥90% of label claim
Microbial Load (TPC, yeast/mold)	Safety and spoilage check	Plate count / ATP assay	<10 CFU/mL
Preservative / Sweetener Validation	Compliance confirmation	HPLC / LC-MS	Within ±10% of label
Heavy Metals	Safety	ICP-MS	As per Codex / FDA limits

Best Practice:

Link analytical test frequency to batch size, risk level, and distribution temperature, higher-risk formulations (low acid, high protein) require more frequent verification.

29.3 Incoming Raw Material Quality Standards

Quality begins with raw materials. Evaluating incoming ingredients for purity, identity, potency, and microbial load prevents downstream variability and protects product integrity. Supplier specifications, certificates of analysis, and standardized acceptance criteria ensure that only qualified materials enter the production stream.

Each ingredient must be verified before use, using supplier CoAs and internal checks.

Table 29.3: Incoming Raw Material Quality Standards and Acceptance Criteria

Material Type	QC Check	Acceptance Criteria
Water	Conductivity, microbiology	<500 μS/cm, TPC <10 CFU/mL
Electrolyte salts	Purity, absence of heavy metals	≥99% purity, Pb <0.1 ppm

Sweeteners	Identity, hygroscopicity	Meets USP/FCC monograph
Natural flavors / colors	Solubility, odor check	No phase separation; clean aroma
Botanical extracts	Identity, active marker	±10% of standard active
Packaging	Seal integrity, migration test	Pass vacuum test, no odor transfer

Supplier Management Tip:

Approve at least two qualified suppliers for each high-volume raw material to mitigate variability and supply disruptions.

29.4 In-Process Quality Monitoring

Real-time monitoring during mixing, heating, homogenization, filling, and cooling allows teams to detect deviations before they affect finished product performance. Tracking key indicators, such as temperature profiles, pH drift, oxygen pickup, and homogenization efficiency, keeps the process aligned with formulation intent and manufacturing specifications.

QC checkpoints should be distributed across the production line to catch deviations early.

Table 29.4: In-Process Quality Monitoring: Critical Control Points and Frequency

Stage	Key Control Parameter	Typical Frequency
Mixing / Blending	pH, Brix, visual clarity	Every batch
Filtration / Deaeration	DO, turbidity	Every batch
Homogenization	Pressure, temperature	Hourly
Pasteurization / HTST	Time–temperature profile	Continuous logging
Filling / Sealing	Headspace O_2, torque	Every 30 min
Packaging	Visual integrity, label check	Every 1,000 units

Formulator’s Insight:

Establish “go/no-go” ranges for every critical control point (CCP). When a variable moves outside range, production should pause for immediate corrective action.

29.5 Finished Product Release Testing

Before a product is cleared for market, release testing verifies that all physical, chemical, microbiological, and sensory parameters fall within specification. These assessments confirm stability, safety, and consistency across the entire batch. Release testing serves as the final checkpoint that protects both brand and consumer.

Every batch must be verified against release specifications before distribution.

Table 29.5: Finished Product Release Testing Protocol

Test	Specification	Method
Appearance	Clear, uniform color, no sediment	Visual & optical test
Odor / Flavor	No off-notes	Sensory panel (trained, n ≥ 5)
pH	Within target ±0.05	pH meter
Brix	Within target ±0.2	Refractometer
Conductivity	Within target ±5%	Conductivity meter
Microbiology	<10 CFU/mL	Plate count
Vitamin C (if added)	≥90% of claim	HPLC assay
Stability marker	No phase separation or precipitation	14-day hold test

Sensory Validation:

Conduct sensory comparison with reference standard (first validated batch) under controlled lighting and temperature (22 ± 1 °C).

29.6 Shelf-Life Validation Protocols

Shelf-life validation measures how the product evolves over time under real and accelerated conditions. Tracking color stability, viscosity changes, nutrient degradation, sensory drift, and microbial integrity allows developers to determine true shelf life and identify early signs of instability. Robust validation supports accurate dating and long-term reliability.

Shelf life is a function of product, packaging, and process. Validation ensures the declared date reflects real-world stability.

Two-Tier Validation System:

1. Accelerated Shelf-Life Test (ASLT)
    - Purpose: Predict degradation kinetics.
    - Condition: 40 °C / 75% RH for 3 months.
    - Goal: No microbial growth; <10% nutrient loss; color ΔE <2.
    - Model: Arrhenius equation to extrapolate real-time shelf life.
2. Real-Time Stability Test
    - Purpose: Confirm actual shelf life at ambient temperature.
    - Condition: 25 °C ± 2 °C; 12 months or until failure.
    - Monitoring: pH, color, nutrient retention, microbial load, sensory attributes.

Formulator's Insight:

Shelf life is not a fixed number; it's a performance window. Define it based on the earliest failing parameter (flavor, vitamin, clarity, or microbial).

29.7 Microbiological Quality Control

Microbiological QC safeguards against spoilage organisms and pathogenic contamination throughout production and storage. Monitoring total plate count, yeast and mold, coliforms, and targeted pathogens ensures beverages remain safe and shelf-stable. Effective microbial control is essential for both regulatory compliance and consumer trust.

Even low-pH beverages can host spoilage organisms. Routine microbiological monitoring prevents recalls.

Table 29.7: Microbiological Quality Control: Target Organisms, Limits, and Testing Methods

Organism Type	Test	Limit	Typical Source
Total Plate Count (TPC)	Plate count agar	<10 CFU/mL	General contamination
Yeasts and Molds	PDA agar	<10 CFU/mL	Sugar, flavor, or air
Lactic Acid Bacteria	MRS agar	<1 CFU/mL	Raw materials
Coliforms / E. coli	MacConkey agar	0 CFU/mL	Water or equipment
Alicyclobacillus spp.	Heat-resistant test	0 CFU/mL	Fruit-based drinks

Sanitation Control:

Perform ATP swab testing on fillers, tanks, and pipelines daily. A pass value is typically <100 RLU (Relative Light Units).

29.8 Stability and Compatibility Testing

Stability and compatibility tests evaluate how ingredients behave together under thermal, mechanical, oxidative, and pH stress. These assessments reveal risks such as precipitation, phase separation, nutrient loss, and flavor interactions. Understanding these dynamics helps refine formulations and protect long-term product quality.

Ingredient–Package Compatibility:

Ensure no migration, corrosion, or flavor adsorption occurs during shelf life.

Table 29.8: Stability and Compatibility Testing Matrix: Ingredient–Packaging Interactions

Test	Purpose	Duration	Acceptance
Migration Test (GC-MS)	Check plasticizer or flavor leaching	6 months	Below regulatory limits
Scalping Test	Measure aroma absorption by PET	3 months	<10% aroma loss
Color Stability (ΔE)	Monitor pigment degradation	3–12 months	<2 units
Vitamin Retention	Nutrient stability	12 months	≥85% retention

Packaging Insight:

Always test packaging and formulation as a system, not separately. Oxygen transmission or acidity can shift dramatically under real product pH.

29.9 Quality Documentation and Traceability

Documentation, batch records, QC logs, deviation reports, and ingredient traceability systems, creates transparency and accountability across the manufacturing chain. Strong documentation ensures regulatory readiness, supports audits, and enables rapid response to quality challenges. Traceability safeguards both consumers and manufacturers.

Table 29.9: Quality Documentation and Traceability Requirements

Document Type	Function	Required Retention
Batch Production Record (BPR)	Captures formulation, process, and QC results	3–5 years

Certificate of Analysis (CoA)	Verifies ingredient compliance	3 years
Deviation & Corrective Action Report (CAPA)	Records nonconformance handling	3 years
Stability Report	Demonstrates shelf-life claim	Product lifetime
HACCP Records	Critical control point verification	3 years minimum
Sensory Evaluation Sheets	Consistency validation	1 year

Digital QC Integration:

Adopt LIMS (Laboratory Information Management System) or ERP-based quality modules to automate tracking, alerts, and audit trails.

29.10 Continuous Improvement Framework

Quality control becomes most powerful when paired with a continuous improvement mindset. Trend analysis, corrective actions, root-cause investigations, and cross-functional reviews help elevate performance over time. A structured improvement framework strengthens product consistency, reduces waste, and enhances operational resilience.

A mature QC system evolves through data-driven continuous improvement.

Table 29.10: Continuous Improvement Framework: Tools for Process and Product Refinement

Tool	Function	Example Use
Statistical Process Control (SPC)	Detects trends and drift	Monitor pH or Brix variation
Root Cause Analysis (RCA)	Investigates deviations	Off-flavor or pH drop

Failure Mode and Effects Analysis (FMEA)	Preventive risk mapping	Packaging O_2 ingress
Sensory Trending	Monitors consumer perception shifts	Flavor fatigue over time

Formulator's Insight:

Every QC metric is a feedback signal to R&D. Track deviations to refine formulations, not just to pass audits.

29.11 Example - Quality Control Specification Summary

A QC specification summary provides a practical snapshot of the parameters, methods, and acceptance ranges required for a functional beverage. This example illustrates how analytical metrics, microbial limits, and sensory standards are integrated into a cohesive quality blueprint for production teams.

Table 29.11: Quality Control Specification Summary: Benchmark Metrics and Testing Frequency

Parameter	Specification	Test Method	Frequency
pH	3.8 ± 0.05	Calibrated pH meter	Batch
Brix	3.5 ± 0.2	Digital refractometer	Batch
Clarity	<1.5 NTU	Nephelometer	Batch
Vitamin C retention	≥90% at 6 months	HPLC	Quarterly
Total plate count	<10 CFU/mL	Plate count	Batch
Dissolved oxygen	<0.5 mg/L	DO sensor	Batch
Color ΔE	<2	Spectrophotometer	Quarterly
Headspace O_2	<0.5%	Analyzer	Batch

Packaging seal	Pass vacuum test	Physical	Every 1,000 units

29.12 Formulation Insight

Formulation insight emerges when QC data is used to refine ingredient selection, processing conditions, and stability strategies. Recognizing which variables consistently influence performance allows developers to adjust formulations with precision. These insights transform quality control from a checkpoint into a powerful tool for product optimization.

- Quality control is not inspection; it's a continuous dialogue between R&D, production, and consumers.
- Always validate process parameters and formulation under the same QC umbrella, the two are inseparable.
- Real-time analytics (inline pH, Brix, O_2) reduce waste and improve precision dramatically.
- A strong QC culture prevents defects instead of detecting them.
- In beverage manufacturing, consistency is the silent promise your brand delivers every time.

30. Sustainability, Sourcing, and Circular Design in Functional Beverages

Sustainability in beverage formulation is no longer optional, it is a strategic differentiator and moral imperative.
Modern consumers expect hydration and nutrition products not only to support personal health but also to respect planetary boundaries.
In this chapter, we explore how to integrate sustainable sourcing, resource efficiency, and circular packaging design into the entire lifecycle of functional electrolyte and nutrition beverages, from raw materials to post-consumer recovery.

30.1 The Sustainability Imperative in Beverage Development

Sustainability has become a core design requirement in modern beverage development, shaping choices around ingredients, processing, packaging, and distribution. Meeting environmental expectations is no longer optional, consumers, retailers, and regulators increasingly demand transparency and measurable impact reduction. Understanding this imperative ensures that functional beverages remain competitive, responsible, and aligned with global sustainability goals.

The global beverage industry contributes significantly to:

- Water usage: up to 3 L of water per 1 L of beverage produced.
- Carbon emissions: from energy-intensive processing and logistics.
- Packaging waste: plastics, multilayer films, and residual contamination.

Sustainability therefore requires a systemic approach, balancing three interdependent pillars:

Table 30.1: Sustainability Pillars in Functional Beverage Development

Pillar	Focus	Beverage Relevance

Environmental	Water, energy, carbon, packaging	Core production footprint
Social	Ethical sourcing, fair labor	Ingredient supply chain
Economic	Efficiency, circular economy	Long-term viability and resilience

Formulator's Insight:

The cleanest label means little if the product's footprint is dirty.
Designing for sustainability begins at the formulation table, not the marketing department.

30.2 Sustainable Ingredient Sourcing and Transparency

Sustainable ingredient sourcing prioritizes traceability, regenerative agriculture, responsible extraction, and reduced environmental footprints. Transparency strengthens consumer trust and ensures that each ingredient aligns with ethical and ecological standards. Integrating sustainability into sourcing decisions enhances both supply-chain resilience and brand integrity.

Sustainable beverage development starts with responsible raw materials, traceable, renewable, and ethically produced.

Table 30.2: Sustainable Ingredient Sourcing Framework and Certification Pathways

Ingredient Type	**Sustainability Consideration**	**Practical Action**
Water	Source, usage, and local impact	Use closed-loop systems, condensate recovery
Botanical Extracts	Biodiversity preservation	Source from certified sustainable farms
Sweeteners	Land and energy intensity	Prefer low-carbon, high-yield crops (stevia, monk fruit)

Proteins / Collagen	Animal welfare, traceability	Use certified, hydrolyzed, low-impact sources
Electrolyte Salts	Mining and refining footprint	Use purified, low-heavy-metal grades
Flavors / Colors	Extraction solvents and yield	Choose CO_2-extracted or water-based processes

Certification Pathways:

- Fair Trade™ (ethical labor, fair pricing)
- Rainforest Alliance™ (sustainable agriculture)
- Non-GMO Project Verified™
- USDA Organic / EU Organic (verified input purity)

Transparency Tip:

Traceability platforms (e.g., blockchain-based or ERP-integrated) are increasingly used to map ingredient journeys, from field to factory.

30.3 Water Stewardship and Efficiency

Water stewardship focuses on reducing usage, optimizing processing efficiency, and protecting water quality across the production cycle. As hydration beverages rely heavily on water, responsible management becomes a defining factor in sustainability performance. Efficient water practices help minimize environmental impact while supporting long-term operational viability.

Water is both ingredient and resource.

Sustainable operations target net water neutrality through reuse, recovery, and optimization.

Table 30.3: Water Stewardship Strategies and Efficiency Metrics in Beverage Manufacturing

Strategy	**Description**	**Impact**

Water Reuse / Recycling	Reclaim rinse and cooling water	20–30% usage reduction
Closed-Loop CIP Systems	Clean-in-place with re-filtered water	Lower chemical + water consumption
Condensate Recovery	Collect steam condensate for pre-heating	Reduces boiler feed demand
Smart Flow Monitoring	Inline meters for each process step	Real-time conservation data
Source Water Mapping	Assess local watershed risk	Supports sustainability audits

Formulator's Insight:

Formulate beverages that require less rinse water, stable emulsions and low-foaming systems simplify line cleaning and reduce water waste.

30.4 Energy and Carbon Reduction Strategies

Energy consumption and carbon emissions are major contributors to a beverage's environmental footprint. Strategies such as process optimization, renewable energy use, heat-recovery systems, and improved line efficiency help reduce greenhouse gas output. Lowering energy intensity enhances sustainability while often improving production economics.

The beverage sector's carbon intensity arises from heating, cooling, and packaging. Each step offers measurable reduction opportunities.

Table 30.4: Energy and Carbon Reduction Opportunities Across Beverage Production Stages

Process Area	Reduction Strategy	Typical Impact
Pasteurization / Thermal Processing	Optimize dwell time, regenerative heat exchangers	-15–25% energy

Cooling & Storage	Use glycol chillers, solar cooling loops	-10–20% energy
Compressed Air Systems	Leak audits and variable frequency drives	-10% energy
Lighting & Facility	LED retrofits, occupancy sensors	-5–10% energy
Logistics	Route optimization, lightweight packaging	-10–30% CO_2

Renewable Integration:

Install solar PV or purchase renewable energy credits (RECs) to offset production emissions.

Carbon Footprint Benchmark:

Typical isotonic beverage = 0.4–0.6 kg CO_2e per 500 mL bottle.
Sustainable systems can reduce this to <0.25 kg CO_2e.

30.5 Sustainable Packaging and Circular Design

Packaging plays a decisive role in environmental impact, making material selection and recyclability central to sustainable beverage design. Lightweight structures, recycled content, refillable models, and circular packaging systems all support waste reduction and resource efficiency. Aligning packaging with circular design principles strengthens both sustainability and consumer appeal.

Packaging is the most visible sustainability touchpoint for consumers.
Circular design focuses on reduce, reuse, recycle, and increasingly regenerate.

Table 30.5: Sustainable Packaging Matrix: Recyclability, Circular Design, and Innovation Readiness

Packaging Type	Recyclability	Circular Design Option
rPET (Recycled PET)	High	Use 25–100% rPET content

Bio-PET	High	Plant-derived resin (sugarcane-based)
Aluminum Can	Very high	Infinite recyclability
Glass Bottle (Returnable)	High	Deposit return systems
Paper-Based Carton	Moderate	Recycled fibers, plant-based liners
Flexible Pouches	Low–Medium	Mono-material PE films, recycle-ready design

Innovation Trends:

- Mono-material caps and labels for full recyclability.
- Digital watermarking for automated sorting.
- Plant-based coatings replacing petroleum laminates.
- Refill & concentrate systems for circular logistics.

Formulator's Insight:

Packaging selection should harmonize with oxygen and light sensitivity, a product that spoils early is never sustainable, regardless of packaging origin.

30.6 Circular Economy in Beverage Production

A circular economy approach reimagines waste as a resource, using strategies such as by-product valorization, material recovery, and closed-loop packaging systems. Applying circular principles reduces environmental burden and increases operational efficiency. This mindset encourages innovation across the entire beverage lifecycle.

A circular economy redefines "waste" as a resource.
Functional beverage manufacturers can adopt circularity at multiple levels:

Table 30.6: Circular Economy Practices in Functional Beverage Production

Circular Practice	Application	Example

By-product valorization	Convert waste streams into usable materials	Fruit pulp → fiber additives
Returnable packaging	Consumer reuse loops	Glass or aluminum refill programs
Renewable input sourcing	Bio-based polymers	PLA coatings, starch films
Nutrient loop optimization	Residuals to feed/fertilizer	Spent botanical material composting
Energy recovery	Anaerobic digestion of organic waste	Generates biogas for steam systems

Industrial Symbiosis Example:

A beverage plant's fruit peel waste can feed a neighboring biogas unit, powering the same plant's pasteurization boiler.

30.7 Sustainable Logistics and Distribution

Transportation and distribution significantly influence carbon footprint, especially for high-volume beverages. Route optimization, lightweight packaging, reduced transportation distances, and alternative fuels all improve sustainability performance. Designing logistics with environmental impact in mind allows brands to reduce both emissions and cost.

Post-production sustainability depends on packaging weight, shelf stability, and regional distribution models.

Table 30.7: Sustainable Logistics and Distribution Optimization Strategies

Strategy	Action	Outcome
Lightweighting	Reduce PET wall thickness by 10–20%	5–10% CO_2 reduction

Regional Production Hubs	Produce near major markets	Cuts freight distance
Cold Chain Optimization	Use phase-change materials	Reduces energy waste
Bulk Concentrate Shipping	Reconstitute at destination	Major logistics savings
Eco-Labeling and QR Transparency	Disclose footprint to consumers	Builds brand trust

30.8 Social Responsibility and Ethical Sourcing

Sustainability extends beyond environmental metrics to include fair labor practices, community support, supplier ethics, and equitable value chains. Ethical sourcing reinforces brand credibility and ensures that beverage development contributes positively to the communities connected to its ingredients. Strong social responsibility strengthens long-term sustainability.

Sustainability also encompasses people, the human chain behind every ingredient.

Table 30.8: Social Responsibility and Ethical Sourcing Implementation Framework

Dimension	**Key Practice**	**Implementation Example**
Fair Labor	Ensure safe conditions and fair wages	Verified supplier audits
Community Investment	Support local farmers / projects	"One Bottle = One Tree" campaigns
Diversity and Inclusion	Encourage inclusive hiring	Workforce training and equal opportunity
Transparency	Disclose sourcing origins	QR-linked supplier maps

Formulator's Insight:

Communicate sustainability stories honestly, evidence-based storytelling resonates more than abstract "eco" claims.

30.9 Measuring and Reporting Sustainability Performance

Accurate measurement, through lifecycle assessments, carbon accounting, water-use audits, and verified impact metrics, creates transparency and accountability. Reporting frameworks translate these metrics into understandable disclosures for regulators, retailers, and consumers. Effective measurement ensures progress is both real and demonstrable.

You can't improve what you don't measure.

Key metrics and reporting standards provide credibility and guidance.

Table 30.9: Sustainability Performance Metrics and Global Reporting Standards

Metric	Unit	Measurement Method	Reference Framework
Carbon Footprint (GHG)	kg CO_2e/L beverage	LCA software, Scope 1–3	GHG Protocol / ISO 14067
Water Footprint	L/L product	Direct + indirect water usage	ISO 14046
Energy Intensity	kWh/L beverage	Utility metering	ISO 50001
Waste Diversion Rate	% waste recycled or recovered	Mass balance	Zero Waste Certification
Packaging Recyclability Index	% recyclable content	Packaging audit	Ellen MacArthur Foundation CE100

Reporting Systems:

- Global Reporting Initiative (GRI)
- Carbon Disclosure Project (CDP)
- Sustainability Accounting Standards Board (SASB)

30.10 Example: Sustainable Hydration Beverage Lifecycle Snapshot

A lifecycle snapshot of a sustainable hydration beverage demonstrates how ingredient sourcing, processing efficiency, packaging choices, and distribution strategies collectively shape environmental impact. This example highlights the interconnected decisions required to create a product that is both functional and environmentally responsible.

Table 30.10: Sustainable Hydration Beverage Lifecycle Snapshot: From Farm to End-of-Life

Lifecycle Stage	Sustainability Practice	Impact
Ingredient Sourcing	Stevia from Rainforest Alliance farms	Ethical labor, biodiversity protection
Formulation	Low-energy extraction, natural colors	Reduced chemical inputs
Processing	HTST with heat recovery loop	25% energy savings
Packaging	100% rPET amber bottle	60% less virgin plastic
Distribution	Regional bottling hub	15% CO_2 reduction
End-of-Life	Recyclable label and cap	Circular material flow

Result:

~40% lower carbon footprint and 30% water savings versus baseline isotonic beverage.

30.11 Formulation Insight

Insight emerges when sustainability considerations are integrated into early formulation stages rather than applied at the end. Recognizing how ingredient choices, processing methods, packaging materials, and supply-chain decisions influence environmental performance allows developers to design beverages that deliver functionality with a reduced ecological footprint.

- Sustainability is an ingredient, measurable, functional, and marketable.
- Align every process step with ISO 14001 environmental management principles.
- Favor modular, efficient formulations, fewer SKUs, fewer raw materials, smaller footprint.
- A sustainable beverage protects both human hydration and planetary regeneration.
- Transparency and data-backed metrics will define the next era of beverage innovation.

31. Formulation Prototyping and Pilot-Scale Implementation

The journey from laboratory innovation to commercial success depends on how well a formulation behaves under real-world manufacturing conditions. Pilot-scale implementation is the bridge that tests whether a beverage's chemistry, stability, and sensory properties can survive the stress of scaling, from milliliters to thousands of liters, without compromising quality.

This chapter provides a structured roadmap for designing, executing, and evaluating pilot-scale trials for electrolyte and nutrition beverages.

31.1 The Purpose of Pilot-Scale Formulation

Pilot-scale formulation bridges the gap between bench development and full-scale manufacturing, revealing how ingredients, equipment, and processing conditions behave under realistic production constraints. This stage validates stability, sensory performance, and manufacturability before significant resources are committed. Understanding its purpose ensures a smooth transition from concept to commercial-ready beverage.

Pilot trials serve as the proof-of-concept for production reality. They confirm that the formulation performs consistently under the same mechanical, thermal, and filling stresses it will experience at industrial scale.

Core Objectives:

1. Validate process–formulation compatibility (mixing, homogenization, heat load).
2. Verify stability and clarity post-processing.
3. Optimize ingredient order of addition and solubility.
4. Identify critical control points (CCPs) for full-scale production.
5. Generate data for regulatory, QC, and sensory validation.

Formulator's Insight:

A successful lab prototype answers "Can this work?"
A successful pilot batch answers "Will it work every time?"

31.2 Scale-Up Strategy and Batch Planning

Effective scale-up requires deliberate planning around batch size, equipment capability, process sequence, and ingredient handling. Strategic alignment between formulation targets and manufacturing realities minimizes variability and production risk. Strong batch planning creates a predictable path from pilot runs to reliable, repeatable commercial output.

Scaling is not linear, physical dynamics (shear, heat transfer, oxygen exposure) change dramatically with batch size.

Table 31.2: Scale-Up Strategy Framework: Batch Sizes, Goals, and Critical Focus Areas

Stage	Batch Size	Objective	Key Focus
Bench	100 mL – 5 L	Proof-of-concept	Ingredient ratio, solubility
Pilot	20 – 500 L	Process simulation	Shear, heat, filling validation
Industrial	≥1,000 L	Production readiness	Efficiency, reproducibility

Critical Scale-Up Rule:

Always scale by energy input per volume (kJ/L), not just by mixing time or speed. If shear energy or residence time changes, so does flavor release and emulsion stability.

31.3 Ingredient Addition Sequence Optimization

The order in which ingredients are added influences solubility, dispersibility, pH behavior, and emulsion or suspension stability. Optimizing the addition sequence prevents precipitation, foaming, and processing inefficiencies. Mastery of sequencing ensures each component performs as intended as batch size increases.

Order of addition determines solubility, pH stability, and final clarity.

Table 31.3: Optimized Ingredient Addition Sequence for Electrolyte and Functional Beverages

Step	Ingredient	Processing Condition	Rationale
1	Water	Start at 25–30 °C	Medium for dissolution
2	Electrolyte salts	Add gradually under agitation	Avoid local supersaturation
3	Acids / Buffers	Adjust pH to 3.8–4.0	Pre-stabilize environment
4	Carbohydrates / Sweeteners	Add at moderate shear	Ensure full dissolution
5	Functional actives (vitamins, amino acids)	Cool to <40 °C	Prevent heat degradation
6	Flavors / Colors	Add under low shear	Preserve aroma
7	Stabilizers (gums, pectin)	Disperse pre-hydrated	Avoid clumping
8	Antioxidants / Preservatives	Add last	Protect other ingredients from oxidation

Formulator's Insight:

Create a master batch sheet specifying order, temperature, and mixing time for every addition, it prevents 90% of scale-up failures.

31.4 Mixing, Shear, and Homogenization Control

Controlling mixing intensity, shear, and homogenization settings is critical for achieving target droplet size, uniform texture, and stable emulsions at pilot scale.

Equipment differences can significantly alter emulsion behavior and physical stability. Fine-tuning these parameters ensures smooth, consistent performance in larger production environments.

Mixing efficiency affects particle dispersion, viscosity, and mouthfeel.

Table 31.4: Mixing, Shear, and Homogenization Parameters Across Scale Levels

Parameter	Lab Scale	Pilot Scale	Control Strategy
Impeller speed	300–600 rpm	100–200 rpm	Match tip speed, not rpm
Shear rate	High	Moderate	Adjust viscosity via stabilizer, not speed
Homogenization pressure	60–100 bar	100–120 bar	Maintain droplet <1 μm
Mixing time	10–20 min	20–45 min	Scale by tank geometry

Equipment Scaling Tip:

Maintain Reynolds number similarity (Re) to ensure equivalent mixing regimes (turbulent or laminar) across scales.

31.5 Thermal Treatment and Cooling Management

Thermal processing at pilot scale, whether pasteurization, hot-fill, or controlled heating, tests how the formulation responds to real equipment and heat-load dynamics. Cooling profiles and hold times must be managed to preserve flavor, nutrient integrity, and stability. Understanding these thermal behaviors ensures the product remains safe and sensorially aligned.

Functional ingredients like vitamins, flavors, and plant extracts are temperature-sensitive.
Pilot trials help determine the minimum viable heat load that guarantees safety without sensory loss.

Table 31.5: Thermal Treatment Protocols and Stability Validation Focus by Beverage Type

Treatment	Typical Condition	Validation Focus
HTST Pasteurization	80–85 °C, 20–30 s	pH drift, color ΔE, flavor
Hot-Fill	90–95 °C, 30 s	Flavor degradation
UHT (Protein Drinks)	140 °C, 2–4 s	Protein aggregation
HPP (Cold-Pressed)	600 MPa, 3 min	Shelf-life confirmation

Cooling Curve Optimization:

Cool to <25 °C within 2 minutes post-heat treatment to minimize vitamin and flavor loss. Use counterflow chillers or glycol plates.

31.6 Filtration and Clarification

Filtration at pilot scale confirms how suspended particles, flavor oils, gums, and colorants interact under real processing conditions. Clarification steps help refine appearance, remove insoluble materials, and protect downstream equipment. Successful filtration ensures visual consistency and long-term stability.

Filtration ensures visual clarity, microbial reduction, and removal of insoluble fines.

Table 31.6: Filtration and Clarification Stages for Functional Beverage Systems

Filtration Stage	Type	Pore Size	Purpose
Coarse Filtration	Cartridge	5–10 µm	Remove undissolved solids
Fine Filtration	Depth / membrane	0.45–1.0 µm	Ensure optical clarity
Sterile Filtration	Membrane	0.22 µm	Cold-fill sterilization

Formulator's Insight:

Avoid over-filtration of natural color systems, anthocyanins and carotenoids can adsorb to filter media, causing fading.

31.7 Filling, Headspace, and Packaging Validation

Pilot-scale filling validates fill accuracy, oxygen pickup, headspace control, and packaging compatibility. These factors strongly influence oxidation rates, microbial stability, and shelf-life performance. Controlled validation ensures that packaging and filling decisions support the beverage's intended quality profile.

Packaging conditions at pilot scale must simulate commercial lines to assess:

- Oxygen ingress
- Seal integrity
- Foam formation
- Fill temperature and headspace control

Table 31.7: Filling, Headspace, and Packaging Validation Metrics for Pilot Trials

Parameter	Target	Test
Fill Temperature	<30 °C (cold fill) / 85–90 °C (hot fill)	Inline thermometer
Headspace O_2	<0.5%	Oxygen analyzer
Torque (cap seal)	1.5–2.0 N·m	Torque meter
Closure Integrity	No leakage after 24h invert test	Vacuum chamber

Pilot Tip:

Use amber PET or aluminum for initial trials, they tolerate slight thermal variability better than glass.

31.8 Sensory and Stability Validation at Pilot Scale

Pilot runs provide the first opportunity to assess how processing affects sensory attributes, color, aroma, sweetness balance, and texture. Stability checks under accelerated and real-time conditions confirm whether the formulation withstands thermal stress, oxygen exposure, and mineral interactions. These validations anchor confidence before full-scale production.

Pilot trials must include both instrumental and human sensory validation.

Table 31.8: Sensory and Instrumental Evaluation Parameters at Pilot Scale

Attribute	Instrumental Method	Sensory Evaluation
Color	Spectrophotometer ($\Delta E \leq 2$)	Visual panel ($n \geq 10$)
Aroma	GC–MS headspace analysis	Trained sensory panel
Sweet–Acid Balance	pH + refractometer	Consumer-like perception test
Mouthfeel / Texture	Viscosity, particle size	Descriptive analysis
Aftertaste	Sensory only	Triangle discrimination test

Formulator's Insight:

Always perform triangle testing (3-sample method) between pilot and lab versions, differences reveal process-related flavor shifts invisible to instrumentation.

31.9 Data Capture and Scale-Up Documentation

Comprehensive documentation, processing parameters, environmental conditions, sensory observations, and analytical data, creates a blueprint for successful scale-up. Accurate data capture enables troubleshooting, repeatability, and quality consistency. This record becomes an essential reference for full-scale manufacturing.

A successful pilot program produces detailed process intelligence, not just a batch of beverage.

Data to Capture:

- Batch records (ingredient weights, addition order, temperatures).
- Equipment parameters (RPM, pressure, dwell time).
- Sensorial and analytical results.
- Deviations and corrective actions.
- Stability observations (clarity, sediment, pH drift).

Template Snapshot – Pilot Batch Report:

Table 31.9: Pilot Batch Report – Data Recording Structure

Section	Key Data Points
Batch Info	Date, operator, batch ID
Ingredients	Lot numbers, weights, sources
Process Parameters	Mix speed, temperature, homogenization pressure
Analytical Data	pH, °Bx, clarity, color, DO
Observations	Foam, aroma release, phase separation
Corrective Actions	Adjusted stabilizer or sequence
Outcome	Passed/failed, recommendations

Formulator's Insight:

Documenting pilot data transforms "trial and error" into measurable optimization.

31.10 Transition from Pilot to Full-Scale Production

Moving from pilot to full-scale production requires translating validated parameters into commercial equipment settings, adjusting for shear differences, heat transfer rates, and batch dynamics. Coordinated alignment among R&D, operations, and quality teams ensures that the finished product matches its pilot-scale performance.

When pilot results meet performance specifications, the next step is industrial validation.

Table 31.10: Scale-Up Transition Protocol: From Pilot to Full Production

Transition Step	Action	Key Consideration
Scale Translation	Match mixing energy and flow dynamics	Use scale factors for shear and heat load
Process Control Mapping	Integrate inline pH, O_2, and Brix sensors	Enables real-time adjustments
Equipment Compatibility	Match homogenizer and heat exchanger type	Avoid shear or temperature shock
Operator Training	Translate pilot SOPs into production instructions	Maintain same order of addition
First Commercial Run	Run 80–90% of full volume	Observe filling and packaging line behavior
Post-Run Validation	Full QC and sensory analysis	Must match pilot reference batch

Scale-Up Reality:

Most issues in first industrial runs are mechanical, not chemical, foam, filling pressure, or pump shear can alter perception more than minor formula differences.

31.11 Example: Pilot-Scale Trial – Citrus Electrolyte Hydration Drink

A pilot-scale trial of a citrus electrolyte hydration drink demonstrates how ingredient sequencing, homogenization, thermal load, and filling conditions interact in real production. This example highlights how targeted adjustments enhance clarity, flavor integrity, and stability through the scale-up process.

Objective: Validate flavor balance, stability, and oxygen management for 12-month shelf life.

Table 31.11: Case Study: Pilot-Scale Trial , Citrus Electrolyte Hydration Drink

Process Step	Parameter	Result
Batch Volume	120 L	Pilot scale
Mixing Temperature	30 °C	Stable dissolution
pH (pre-fill)	3.8 ± 0.05	Within target
Homogenization	100 bar, 2 passes	Droplet <1 μm
HTST Pasteurization	82 °C, 25 s	No flavor burn
Cooling	<25 °C within 2 min	Good vitamin retention
Fill Type	Cold-fill with N_2 flush	DO = 0.4 mg/L
Packaging	Amber PET	Stable under 40 °C, 75% RH
3-Month Stability	No color shift, pH drift <0.03	Pass

31.12 Common Pilot-Scale Challenges and Solutions

Pilot scale often exposes issues not seen at bench level, such as incomplete dissolution, flavor shifts, viscosity changes, or emulsion instability. Identifying these challenges and applying targeted solutions strengthens formulation robustness and reduces risk before full-scale launch.

Table 31.12: Common Pilot-Scale Challenges and Corrective Strategies

Issue	Cause	Solution
Foaming during filling	Surfactant or stabilizer imbalance	Add antifoam (0.01%) or reduce shear
Flavor loss	Excessive mixing time / heat	Add aroma post-pasteurization

Sediment formation	Poor dispersion or ionic incompatibility	Adjust gum ratio, pre-hydrate hydrocolloid
Vitamin degradation	Oxygen exposure	Add antioxidant at end, nitrogen flush
Color fading	UV or oxidation	Use amber PET, reduce O_2 ingress

31.13 Formulation Insight

Insight emerges when pilot-scale learnings are used to refine both formulation and process design. Recognizing how ingredient behavior changes with scale, shear, heat, and oxygen exposure leads to more resilient beverage systems. These insights form the foundation for reliable, high-quality commercial production.

- Pilot trials are where science meets machinery, your formulation's first real test of survival.
- Always document why each deviation occurred, not just what happened.
- Treat pilot-scale outcomes as living feedback loops for R&D and QC refinement.
- Scale-up is not replication, it's translation, preserving performance under industrial stress.
- Success is achieved when the beverage tastes identical at 5 liters and at 5,000 liters.

32. Case Studies and Benchmark Formulations

Case studies transform formulation theory into real, reproducible beverage systems.
They show how ingredient interactions, sensory engineering, and process design translate into commercial-ready products with validated stability, flavor, and functionality.

This chapter presents benchmark formulations across several categories of electrolyte and nutrition beverages, each optimized for its performance role, target consumer, and sensory signature.

32.1 Case Study 1: Isotonic Hydration Drink (Citrus Performance Blend)

This isotonic hydration drink illustrates how carefully balanced electrolytes, carbohydrates, acids, and natural citrus flavors combine to deliver rapid fluid uptake and strong sensory appeal. The case study highlights the formulation logic behind osmolality control, sweetness architecture, and clarity management in performance-driven beverages.

Objective: Develop a clear, refreshing isotonic beverage for rehydration during moderate to intense physical activity.

Product Specification Overview:

Table 32.1A: Isotonic Citrus Performance Hydration Drink

Category	Specification
Target Functionality	Rapid hydration, electrolyte replenishment
Caloric Load	25 kcal/100 mL
Osmolality	280–300 mOsm/kg
pH Range	3.6–3.8

Shelf Life	12 months ambient

Formulation:

Table 32.1B: Benchmark Formulation: Isotonic Citrus Performance Hydration Drink

Ingredient	% w/w	Function
Purified water	92.50	Solvent / base
Sucrose	3.00	Energy / isotonic balance
Sodium citrate	0.25	Buffer / electrolyte
Potassium chloride	0.10	Electrolyte balance
Magnesium lactate	0.05	Muscle support
Citric acid (anhydrous)	0.35	pH adjustment
Natural citrus flavor (oil-based)	0.30	Flavor system
Natural color (β-carotene)	0.005	Appearance
Vitamin C (ascorbic acid)	0.02	Antioxidant / functional
Sodium benzoate	0.015	Preservative
Deaerated water (make up)	To 100%	Volume balance

Processing Summary:

- Mix electrolytes and sugar in water at 35 °C.
- Adjust pH to 3.7.
- Add flavor and color under gentle shear.
- HTST pasteurize at 82 °C for 25 s, cool to 25 °C.
- Fill under nitrogen.
- Stability validated for 12 months, $\Delta E < 1.5$, pH drift < 0.05.

Sensory Profile

- Aroma: Fresh lemon-lime top note.
- Taste: Balanced sweet-sour with mineral smoothness.
- Mouthfeel: Clean, light, slightly tangy finish.

Key Learning:

Sodium–potassium ratio of 1.5:1 gives optimal rehydration efficiency without salty aftertaste.

32.2 Case Study 2: Electrolyte + Collagen Beauty Water (Floral Citrus Blend)

An electrolyte + collagen beauty water demonstrates how protein stability, delicate flavor systems, and pH management intersect to create a clean, visually refined functional beverage. This case study explores how floral–citrus aromatics, natural colorants, and collagen compatibility shape both performance and premium positioning.

Objective: Develop a premium, low-calorie, functional hydration beverage targeting skin wellness and daily beauty maintenance.

Product Specification Overview:

Table 32.2A: Electrolyte + Collagen Beauty Water (Floral Citrus Blend)

Specification	Value
Target Consumer	Female, wellness/lifestyle
Caloric Load	<10 kcal/100 mL
Protein Source	Hydrolyzed collagen (Type I/III)
Shelf Life	9 months
Visual Target	Clear, pastel pink hue

Formulation:

Table 32.2B: Benchmark Formulation: Electrolyte + Collagen Beauty Water (Floral Citrus Blend)

Ingredient	% w/w	Function
Purified water	94.60	Base
Collagen peptide (2000 Da)	0.60	Functional protein
Citric acid	0.25	pH control
Sodium citrate	0.15	Buffer
Potassium chloride	0.05	Electrolyte
Magnesium citrate	0.05	Mineral source
Stevia extract (Reb A 97%)	0.015	Natural sweetener
Rose + citrus flavor	0.25	Aroma system
Natural color (red radish extract)	0.005	Tint
Ascorbic acid	0.02	Antioxidant
Preservative (potassium sorbate)	0.015	Shelf-life protection
Deaerated water	To 100%	Balance

Process:

- Dissolve collagen under moderate agitation at 45 °C.
- Add electrolytes, acid, and buffers.
- Cool to < 35 °C; add flavors and color.
- Cold-fill under N_2 flush.

Sensory & Stability:

- Flavor: Floral-citrus, gentle sweetness, clean aftertaste.
- Clarity: < 1.0 NTU after 6 months.

- Vitamin retention: 87% at 9 months.

Key Learning:

Collagen's clarity improves when the pH is maintained between 3.6 and 3.8, minimizing peptide aggregation.

32.3 Case Study 3: Plant Electrolyte Tonic (Coconut + Matcha)

A plant-based electrolyte tonic featuring coconut minerals and matcha highlights the challenges and opportunities of formulating with botanicals, natural pigments, and plant-derived minerals. This example showcases how flavor balancing, antioxidant protection, and clean-label emulsification come together to form a modern, plant-forward hydration system.

Objective: Create a plant-forward hydration tonic combining coconut electrolytes with matcha-derived antioxidants.

Table 32.3A: Functional Design Brief: Plant Electrolyte Tonic (Coconut + Matcha)

Functionality	**Hydration + antioxidant support**
Total Solids	4.5%
Energy	20 kcal/100 mL
Shelf Life	9–12 months
Appearance	Slightly cloudy, natural green tint

Formulation:

Table 32.3B: Benchmark Formulation: Plant Electrolyte Tonic (Coconut + Matcha)

Ingredient	**% w/w**	**Function**
Coconut water concentrate	10.00	Electrolyte base
Purified water	88.60	Base

Matcha extract	0.10	Antioxidant source
Citric acid	0.20	pH adjustment
Sodium citrate	0.10	Buffer
Natural sweetener blend (erythritol + stevia)	0.70	Sweetness balance
Natural flavor (coconut–green tea)	0.20	Aroma
Ascorbic acid	0.05	Antioxidant
Stabilizer (pectin, LM type)	0.05	Suspension aid
Potassium sorbate	0.015	Preservative
Deaerated water	To 100%	Balance

Processing:

- Mix coconut concentrate and water; dissolve electrolytes.
- Add matcha under high shear, then pectin pre-hydrated in warm water.
- HTST at 82 °C for 25 s; homogenize 2 × 80 bar.
- Cold-fill under N_2.

Outcome:

- Stable suspension, natural vegetal color retained.
- $\Delta E < 2$ over 12 months; total polyphenol retention > 80%.

Key Learning:

Use LM-pectin to stabilize natural polyphenols and prevent sedimentation without opacity.

32.4 Case Study 4: Performance Recovery Drink (Electrolytes + BCAA)

This recovery beverage demonstrates how amino acids, electrolytes, sweetness modulators, and aroma systems can be aligned to overcome bitterness and achieve post-exercise functionality. The case study unpacks strategies for taste masking, osmolality management, and nutrient stability in performance-focused formulations.

Objective: Post-workout beverage supporting muscle recovery and rehydration.

Protein Source	Branched-chain amino acids (BCAA 2:1:1)
Calories	35 kcal/100 mL
Shelf Life	12 months
Flavor Profile	Tropical mango-pineapple

Formulation

Table 32.4: Benchmark Formulation: BCAA-Enhanced Performance Recovery Drink

Ingredient	% w/w	Function
Water	90.00	Solvent
Glucose	4.00	Rapid energy
BCAA blend	1.50	Muscle recovery
Sodium citrate	0.30	Buffer
Potassium chloride	0.10	Electrolyte
Citric acid	0.35	Acidity
Natural tropical flavor	0.35	Aroma
β-carotene color	0.005	Visual identity
Sucralose + acesulfame K	0.015	Sweetener

Ascorbic acid	0.02	Antioxidant
Preservative	0.015	Stability

Processing:

- High-shear disperse BCAA into warm water (40 °C).
- Add sugars and salts, then acidify to pH 3.8.
- Cool before adding flavor and color.
- Pasteurize HTST 82 °C / 20 s.

Sensory Notes:

- Mildly sweet–tart with soft amino aftertaste (masked via citrus–mango top notes).
- Clean finish, stable at ambient.

Key Learning:

Adding BCAA before acidification minimizes foaming and solubility loss during processing.

32.5 Case Study 5: Functional Electrolyte Shot (High-Concentration Format)

A functional electrolyte shot explores the challenges of creating a compact, high-strength format with elevated electrolytes, potent flavor actives, and concentrated acidity. This example highlights strategies for managing intensity, achieving rapid stability, and maintaining consumer-friendly sensory balance in small-volume applications.

Objective: Compact 60 mL format delivering electrolyte and vitamin boost for athletes.

Osmolality	550–600 mOsm/kg
Shelf Life	9 months
Target pH	3.5

Formulation

Table 32.5: Benchmark Formulation: Functional Electrolyte Shot (High-Concentration Format)

Ingredient	% w/w	Function
Purified water	74.00	Base
Glucose syrup	15.00	Osmotic energy
Electrolyte mix (NaCl, KCl, Mg lactate)	2.00	Rehydration
Citric acid	0.40	pH control
Sodium citrate	0.20	Buffer
Vitamin C + B-complex	0.05	Functional actives
Natural berry flavor	0.20	Aroma
Natural red color	0.01	Visual appeal
Preservative	0.015	Shelf stability
Deaerated water	To 100%	Volume adjustment

Outcome:

- Stable viscosity (1.2 mPa·s), no crystallization.
- Retains 92% vitamin potency at 6 months.

Key Learning:

Maintain low pH and add vitamins post-cooling to protect heat-sensitive nutrients.

32.6 Comparative Analysis of Formulation Types

Comparing these case studies reveals how different product formats, hydration drinks, beauty waters, botanical tonics, recovery blends, and concentrated shots, require distinct strategies in sweetness design, stability management, ingredient compatibility, and processing. This analysis highlights the unique trade-offs and technical priorities across the functional beverage landscape.

Table 32.6: Comparative Analysis of Benchmark Formulation Types - Performance, Beauty, Plant, and Recovery Systems

Parameter	Isotonic Hydration	Collagen Beauty	Plant Tonic	BCAA Recovery	Electrolyte Shot
Target User	Athletes, daily users	Wellness / lifestyle	Natural / vegan	Sports recovery	Endurance athletes
Calories (per 100 mL)	25	<10	20	35	45
pH	3.7	3.7	3.8	3.8	3.5
Shelf Life (months)	12	9	12	12	9
Clarity	Clear	Clear–tinted	Slight haze	Clear	Clear
Processing Type	HTST	Cold-fill	HTST	HTST	HTST
Sweetening System	Sugar	Stevia	Natural blend	Artificial mix	Sugar + sweeteners
Stabilizers	None	Light	Pectin	None	None

32.7 Key Takeaways Across Case Studies

Key lessons emerge across the examples: stability improves through synergy rather than single fixes; flavor systems must be built holistically; and processing conditions often dictate final performance. These takeaways provide a distilled view of the principles that consistently drive successful functional beverage design.

- Balance function and flavor: Consumers accept mild off-notes if performance benefit is clear and flavor masking is skillful.

- Optimize solubility early: Pre-dissolution of electrolytes and amino acids avoids precipitation later.
- Pilot validation is non-negotiable: Each case required process fine-tuning before scale-up.
- pH is your hidden stabilizer: Every beverage maintained between 3.5 – 3.8 for optimal clarity and safety.
- Formulate for shelf life, not for the lab: Nutrient retention and color integrity define commercial success.

32.8 Formulation Insights

Insight arises when the common patterns across case studies, such as pH control, electrolyte interactions, masking strategies, and packaging influence, are translated into actionable formulation wisdom. These learnings equip developers to navigate diverse product types with confidence and precision, applying foundational principles to new and innovative beverage concepts.

- Case studies show how science and design merge, chemistry dictates functionality, but sensory drives adoption.
- A successful beverage must satisfy three tests: analytical stability, process resilience, and emotional appeal.
- The best formulations tell a story through their ingredient architecture, purposeful, transparent, and data-validated.
- These blueprints can be adapted regionally through flavor localization and regulatory alignment, while maintaining core performance.

33. Advanced Sensory Design, Consumer Validation, and Market Readiness Testing

Formulation science ensures stability, safety, and function, but sensory science ensures loyalty and emotional connection.
A hydration or nutrition drink must not only deliver physiological benefit but also signal freshness, trust, and pleasure at the first sip.

This chapter outlines how to scientifically engineer and validate flavor, aroma, mouthfeel, and color perception to achieve market-ready acceptance.

33.1 The Role of Sensory Science in Functional Beverage Development

Sensory science provides the structured, evidence-based framework needed to translate formulation decisions into predictable consumer experiences. In functional beverages, where electrolytes, acids, botanicals, and proteins can create challenging sensory profiles, sensory science ensures that performance and palatability move in parallel. Understanding this role allows developers to design beverages that deliver both functional impact and intuitive consumer acceptance.

Functional beverages challenge traditional sensory design, electrolytes, acids, proteins, and botanicals all create taste conflicts that must be harmonized.

Goals of Sensory Design:

1. Achieve balanced flavor architecture across sweet, sour, bitter, and salty dimensions.
2. Maintain cross-modal harmony, alignment of taste, aroma, and mouthfeel.
3. Ensure emotional coherence, product's sensory cues must match its claimed benefit (e.g., "energizing," "calming").
4. Predict consumer preference before commercial launch.

Formulator's Insight:

Consumers forgive mild aftertastes but never confusion, clarity of flavor message (what it tastes like and what it stands for) is more important than technical perfection.

33.2 Building a Sensory Profile Framework

A sensory profile framework defines the target experience across taste, aroma, mouthfeel, aftertaste, and visual attributes. Establishing these benchmarks early guides ingredient selection, flavor architecture, and stability goals. A clear framework becomes the blueprint that aligns formulation work with the intended sensory identity of the beverage.

Each beverage should have a defined sensory fingerprint developed using structured descriptive analysis.

Table 33.2: Sensory Profile Framework: Attribute Categories, Descriptors, and Instrumental Correlates

Attribute Category	Descriptor Examples	Instrumental Correlation
Appearance	Color tone, clarity, brightness	Colorimeter (L*, a*, b*)
Aroma	Citrus, floral, herbal, metallic	GC–MS volatile profiling
Taste	Sweetness, sourness, bitterness, saltiness, umami	Electronic tongue, titration
Mouthfeel	Body, smoothness, viscosity	Rheometer
Aftertaste	Persistence, cleanliness	Time–intensity analysis

Toolset:

- QDA (Quantitative Descriptive Analysis)
- TDS (Temporal Dominance of Sensations)

- Napping® and Projective Mapping for quick visual sensory space mapping.

Data Integration:

Combine sensory scores with instrumental analytics (pH, viscosity, °Brix, DO) to find root causes of sensory drift.

33.3 Designing for Functional Flavor Harmony

Functional flavor harmony arises when sweetness, acidity, aroma, and masking strategies integrate seamlessly with electrolytes, amino acids, botanicals, or vitamins. Designing this harmony requires understanding how functional ingredients shift perception and how compensatory flavor elements can restore balance. Mastery of these interactions produces beverages that taste clean, coherent, and purpose-driven.

Functional drinks contain actives that distort flavor balance (e.g., magnesium = bitter, amino acids = sulfurous).
Sensory engineering neutralizes these effects using masking, counterbalance, and associative cues.

Table 33.3: Functional Flavor Harmony Matrix: Off-Note Sources and Engineering Solutions

Issue	Sensory Perception	Engineering Strategy
Sodium or potassium	Salty / metallic	Use organic salts (citrate, gluconate)
Magnesium or calcium	Bitter / chalky	Blend with tart citrus or berry acids
BCAA or amino acids	Sulfurous, lingering	Add tropical top notes (mango, passionfruit)
Stevia or monk fruit	Delayed sweetness, licorice note	Blend with low-dose sucrose or erythritol

Botanicals (adaptogens)	Earthy, bitter	Pair with green tea or ginger notes
Collagen	Slight meaty / gelatin note	Floral-citrus aroma overlay

Formulator's Insight:

The sweet–acid–aroma triangle defines acceptance:

- Sweet masks salt and bitter.
- Acid amplifies aroma.
- Aroma primes sweetness perception.

33.4 Sensory Mapping and Competitive Benchmarking

Sensory mapping and benchmarking position a beverage within the competitive landscape by comparing its flavor intensity, sweetness curve, acidity, mouthfeel, and aromatic structure to leading products. This approach reveals gaps, opportunities, and differentiation strategies. Mapping ensures that the beverage not only performs well but also fits the expectations of its category.

Before launch, each formulation should be mapped against market competitors to identify sensory white space.

Example: 2D Sensory Map (Citrus Electrolyte Beverages)

- X-axis: Sweetness intensity
- Y-axis: Freshness perception (acid–aroma balance)

Cluster analysis identifies:

- "Overly sweet" mainstream isotonic drinks
- "Flat" low-acid hydration waters
- "Ideal zone" (moderate sweetness, bright acid, natural aroma)

Your product should target the center of sensory preference cluster identified through consumer data (usually moderate intensity, not extremes).

33.5 Advanced Flavor Masking and Enhancement Technologies

Modern masking tools, encapsulation systems, temporal modulators, bitterness blockers, and targeted aroma enhancers, push sensory precision beyond traditional formulation. These technologies help neutralize harsh notes from minerals, amino acids, botanicals, and vitamins while elevating desirable character. Leveraging advanced tools enables cleaner, more refined sensory profiles in complex functional systems.

Table 33.5: Advanced Flavor Masking and Enhancement Technologies

Technique	Mechanism	Example Application
Microencapsulation	Isolates bitter actives	BCAA, caffeine
Lipid-carrier flavor encapsulation	Improves top-note release	Citrus oils
Sweetness modulation compounds	Enhance sucrose perception	Thaumatin, glycosides
Bitter blockers	Inhibit receptor binding	AMP, adenosine derivatives
Cross-modal enhancers	Aroma–taste synergy	Lemon aroma enhancing perceived acidity
Thermal aroma restoration	Post-pasteurization volatile replenishment	Vitamin drinks

Formulator's Insight:

Flavor masking shouldn't "fight" off-notes, it should redirect them into familiar sensory territories (e.g., metallic → "mineral freshness").

33.6 Consumer Validation and Preference Modeling

Consumer validation transforms sensory prototypes into real-world insights by identifying preference drivers, acceptance thresholds, and improvement opportunities. Preference modeling reveals the relationships between sweetness, acidity, flavor type, mouthfeel, and overall liking. These findings ensure the final product resonates with its intended audience.

After internal sensory optimization, validation with target consumers confirms market alignment.

A. Central Location Testing (CLT)

- 100–200 participants, blinded samples.
- 9-point hedonic scale for taste, aroma, color, and aftertaste.
- Ideal mapping: Mean liking ≥ 7/9 across attributes.

B. Home Use Tests (HUT)

- 2–4 week in-home consumption.
- Measures sustained liking and repeat intent.
- Useful for functional beverages (habit-driven).

C. Preference Mapping

Statistical correlation between sensory data (from expert panel) and consumer liking (from CLT/HUT).

Table 33.6: Consumer Validation Protocols: CLT, HUT, and Preference Modeling

Model Type	**Purpose**
Internal Preference Mapping	Identify consumer clusters
External Preference Mapping	Link chemical/sensory variables to liking
Penalty Analysis	Quantify sensory drivers of dislike

Key Insight:

Consumers prefer beverages with moderate sweetness, bright aroma, and smooth finish, extremes polarize acceptance.

33.7 Color and Visual Psychology in Market Readiness

Color sets immediate expectations for flavor, function, and quality. Visual psychology shapes how consumers interpret freshness, strength, and benefit cues before tasting the beverage. Understanding these cues ensures that color, clarity, opacity, and packaging aesthetics reinforce the beverage's functional promise and sensory identity.

Visual cues heavily influence flavor expectation and perceived functionality.

Table 33.7: Color and Visual Psychology Matrix for Functional Beverages

Color Tone	Sensory Expectation	Functional Association
Citrus yellow	Sour, fresh	Hydration, performance
Pale pink	Light, floral	Beauty, relaxation
Green	Natural, detox	Plant, antioxidant
Amber / gold	Warm, energy	Focus, strength
Blue	Cool, refreshing	Electrolyte, cooling

Formulator's Insight:

Clarity = purity perception.
Even "cloudy" natural beverages can be perceived as clean if color is stable and hue is consistent.

33.8 Sensory Stability Testing

Sensory stability testing evaluates how flavor, aroma, mouthfeel, and color evolve throughout shelf life. Oxidation, ingredient interactions, pH drift, and aroma loss all influence long-term sensory performance. Testing allows developers to identify subtle changes early and ensure the beverage remains appealing from production to expiry.

Sensory performance must endure shelf life, not just chemistry.

Table 33.8: Sensory Stability Testing Parameters and Acceptability Thresholds

Parameter	Method	Threshold for Acceptable Drift
Color stability (ΔE)	Spectrophotometer	<2 units at 12 months
Flavor volatility	GC–MS headspace	<15% top-note loss
Sweetness drift	Brix + sensory test	<0.3 °Bx difference
Mouthfeel viscosity	Rheometer	<10% deviation
Aroma intensity (hedonic)	Panel mean	≤ 0.5 scale unit loss

Accelerated Sensory Aging:

40 °C storage, 3 months = ~1 year ambient sensory drift projection.

33.9 Market Readiness Checklist

A market readiness checklist integrates sensory, analytical, regulatory, and packaging considerations to confirm that the beverage is fully prepared for commercial launch. This structured assessment ensures alignment between product experience, category expectations, and brand positioning. Completing the checklist reduces risk and strengthens launch success.

Before commercial release, every beverage should pass the Market Readiness Matrix, aligning product integrity with consumer promise.

Table 33.9: Market Readiness Matrix - Validation Metrics and Pass/Fail Targets for Commercial Launch

Dimension	Validation Metric	Target
Flavor Performance	Consumer liking ≥ 7/9	Pass

Aroma Consistency	Headspace GC profile ±10%	Pass
Color Stability	ΔE < 2 after 6 months	Pass
Nutrient Retention	≥90% at shelf life	Pass
Label Compliance	100% verified	Pass
Packaging Performance	No leak / discoloration	Pass
Consumer Understanding	≥80% correct benefit recall	Pass
Repurchase Intent	≥75%	Pass

33.10 Example: Consumer Validation Snapshot

A consumer validation snapshot illustrates how preference data, hedonic scoring, and sensory diagnostics come together to guide refinements in flavor, sweetness, aroma, and texture. This example demonstrates how consumer insights translate into actionable formulation decisions before scaling to full production.

Product: Electrolyte + Collagen Beauty Water
Testing: 150 female participants, ages 20–40, wellness-oriented consumers

Table 33.10: Consumer Validation Snapshot: Mean Liking, Comments, and Market Readiness Indicators

Attribute	Mean Liking (1–9)	Comments
Taste	8.2	"Light, refreshing, natural"
Aroma	8.4	"Soft floral note is elegant"
Color	8.1	"Looks premium, gentle pink"
Mouthfeel	7.8	"Smooth and easy to drink"
Aftertaste	7.5	"Clean, slightly citrus"
Purchase Intent	82%	Excellent readiness

Result: Proceed to launch; only adjustment suggested, slight increase in aroma intensity to improve differentiation on first sip.

33.11 Formulation Insights

Insight emerges when sensory learnings are fed back into formulation, adjusting sweetness curves, modulating acidity, refining aroma systems, or improving masking strategies. Understanding how sensory data aligns with functional design enables developers to create beverages that perform physiologically while delivering an exceptional consumer experience.

- Sensory design is emotional chemistry; it translates data into desire.
- The most successful beverages don't taste complex; they feel balanced.
- Use sensory drift data as an early-warning system for formulation instability.
- Market readiness is not about perfection, but predictability under variability, shelf life, temperature, and consumer diversity.
- A functional drink succeeds when it tastes like self-care, effortless, trustworthy, and worth repeating.

34. Future Trends and Innovation Pathways in Functional Hydration and Nutrition Drinks

The global beverage market is entering a new phase of convergence: performance, personalization, and planet responsibility.
Consumers now demand drinks that not only hydrate but also optimize, restore, and sustain, all without compromising naturalness, taste, or transparency.

This chapter explores the next-generation technologies, ingredients, and design philosophies shaping the future of electrolyte and functional nutrition beverages.

34.1 Macro Trends Driving the Next Wave

Shifting consumer priorities, advances in nutritional science, and accelerating sustainability demands are reshaping the landscape of functional beverages. These macro trends define the next wave of innovation, guiding how brands approach formulation, processing, and product positioning. Understanding these forces provides a roadmap for designing beverages that meet future expectations.

Table 34.1: Macro Consumer and Market Trends Shaping Next-Generation Hydration Beverages

Trend	Description	Implication for Beverage Innovation
Holistic Wellness	Consumers link hydration to mood, immunity, and energy	Functional ingredients must deliver multi-system benefits
Personalization	Rise of DNA, microbiome, and lifestyle-based nutrition	Modular, data-driven drink systems
Clean & Conscious Labels	Short ingredient lists, recognizable sourcing	Natural actives, transparency, minimal additives

Sustainability & Circularity	Carbon neutrality and regenerative sourcing	Local sourcing, refill models, upcycled ingredients
Convenient Functionality	On-the-go yet purposeful	Concentrates, shots, and powdered reconstitutes
Tech-Enabled Nutrition	Integration with wearables and apps	Smart hydration tracking, adaptive formulations

Formulator's Insight:

Tomorrow's hydration will be data-responsive, tailored to physiology and environment, not just general need.

34.2 Emerging Ingredient Frontiers

New ingredient frontiers, from adaptive botanicals and precision-fermented nutrients to bioactive peptides and advanced electrolyte complexes, are expanding what functional beverages can achieve. These novel inputs offer enhanced efficacy, stability, and bioavailability while opening the door to entirely new benefit categories. Exploring these frontiers helps developers stay ahead of innovation curves.

1. Bioactive Electrolyte Systems

- Next-gen mineral carriers: Gluconates, malates, and amino-acid chelates for higher bioavailability.
- Microencapsulated salts: Controlled release in oral cavity for flavor masking and gradual absorption.
- Smart osmolality design: Adaptive isotonic systems adjusting to exertion level.

2. Cognitive and Mood Modulators

- L-theanine, GABA, saffron extract, and nootropic botanical stacks.
- Designed for mental hydration: focus, calm, and cognitive stamina.

- Integration with electrolyte bases to create "mind-body balancing" drinks.

3. Adaptogenic and Metabolic Botanicals

- Ashwagandha, cordyceps, ginseng, rhodiola, positioned for resilience and recovery.
- Used at physiologically relevant but sensory-safe doses.
- Often paired with natural acids (e.g., malic, citric) for flavor synergy.

4. Natural Sweetness 3.0

- Blends of stevia glycosides (Reb M/D) + monk fruit for clean taste.
- Enzymatically modified stevia for sucrose-like onset.
- Future frontier: precision fermentation sweeteners, identical to natural, lower impact.

5. Microbiome-Friendly Electrolytes

- Synbiotic hydration drinks: combine minerals with prebiotics (inulin, FOS) and postbiotics for gut support.
- Potential crossover category: hydration + digestion harmony.

34.3 Technological Innovation in Processing and Stability

Next-generation processing technologies, smart stabilization tools, and integrated digital monitoring systems are revolutionizing how beverages are produced and protected. These innovations improve nutrient retention, reduce energy inputs, enhance microbial safety, and broaden formulation flexibility. Mastery of emerging technology enables cleaner, more resilient, and more efficient product development.

Table 34.3: Advanced Processing and Stability Technologies on the Horizon

Technology	Innovation	Impact

Low-energy HTST systems	Regenerative heat recovery	-25% carbon footprint
High-pressure homogenization (HPH)	Uniform microemulsions	Better flavor and color stability
Nanoencapsulation	Protects flavors, vitamins, and botanicals	Enhanced shelf life and taste
Electrochemical sterilization	Non-thermal, ion-based	No flavor loss, lower energy
Digital twin processing	Virtual model of beverage line	Predicts process deviations before they occur

Formulator's Insight:

Innovation now lives at the intersection of biology, data, and process engineering. Tomorrow's beverage plant will look more like a biotech lab than a bottling hall.

34.4 Personalized Hydration Systems

Personalized hydration is shifting beverages from generalized solutions to individualized performance tools informed by biomarkers, activity data, and physiological needs. Modular formulations, adaptive dosing, and connected delivery systems are redefining how hydration is tailored. This emerging paradigm moves functional beverages closer to precision nutrition.

Smart hydration is the emerging frontier linking technology with physiology.

Table 34.4: Personalized Hydration Delivery Models and Implementation Pathways

Model	Mechanism	Example
Wearable-integrated systems	Sensors track sweat rate and electrolyte loss	Adaptive rehydration via app

AI-driven formulation hubs	Cloud-based blend customization	Dispensers that tailor mixes per user
Cartridge-based home systems	Smart pods with data-linked profiles	Personalized hydration-as-a-service
Micro-dosed concentrates	Ultra-concentrated actives with AI guidance	Portable, sustainable, customizable

Market Direction:

Next-gen brands are not selling "a drink," but a hydration ecosystem, data, hardware, and formulation connected by personalization algorithms.

34.5 Functional Beverage Architecture: Layered Benefits

Modern consumers seek multifunctional systems that support hydration, energy, cognition, immunity, and recovery simultaneously. Designing beverages with layered benefits requires building synergistic ingredient architectures that deliver multiple outcomes without compromising sensory or stability. This approach signals a new era of integrated functional performance.

Consumers are moving away from single-benefit drinks toward stacked functionality.
Tomorrow's electrolyte beverages will deliver synergistic clusters of benefit:

Table 34.5: Functional Beverage Architecture: Layered Benefit Clusters for Use-Case Targeting

Functional Cluster	Example Ingredient Stack	Benefit Focus
Hydration + Focus	Electrolytes + L-theanine + guayusa extract	Clean alertness
Hydration + Calm	Magnesium + chamomile + lemon balm	Stress balance

Hydration + Skin Health	Collagen + vitamin C + hyaluronic acid	Beauty from within
Hydration + Energy	Sodium + caffeine microdose + B-vitamins	Active performance
Hydration + Immunity	Zinc + acerola + polyphenols	Daily defense

Formulator's Insight:

Consumers are building micro-rituals, morning energy hydration, midday clarity tonic, evening calm water.
Your formulation architecture should anticipate use occasions, not just nutrients.

34.6 The Rise of "Bio-Hydration" Platforms

Bio-hydration platforms integrate bioactive compounds, microbiome-supportive components, targeted electrolytes, and metabolic modulators to elevate hydration beyond fluid replacement. These systems emphasize cellular-level efficiency, recovery enhancement, and metabolic support. Their emergence reflects the fusion of hydration science with advanced functional nutrition.

A new research frontier explores cellular hydration, not just water replacement, but membrane-level water dynamics.

Table 34.6: "Bio-Hydration" Platforms: Cellular-Level Water Dynamics and Emerging Science

Innovation Area	Description	Future Potential
Aquaporin Modulators	Nutrients that influence water channel proteins	Enhanced cellular uptake
Electrolyte–polyol synergy	Optimized osmotic balance	Slower dehydration during heat exposure
Structured water technologies	Clustered hydrogen-bond networks (still under study)	Bioavailability enhancement

Emerging Research Direction:

Exploring how plant polyols, amino acids, and low-molecular polysaccharides affect intracellular hydration kinetics.

34.7 Sustainable and Circular Beverage Systems

Future beverages must operate within fully circular systems, minimizing waste, maximizing resource efficiency, and utilizing regenerative ingredient sourcing. Packaging, processing, and supply-chain design are evolving toward lower-impact models that close loops and reduce emissions. Sustainability becomes inseparable from formulation strategy.

Sustainability innovation is shifting from "less impact" to positive regeneration.

Table 34.7: Regenerative and Circular Beverage System Innovations

Frontier	Innovation	Example
Upcycled Ingredients	Valorized fruit fibers, coffee fruit, spent grain extracts	Reduces waste, adds nutrients
Regenerative Sourcing	Partnering with farms restoring soil and biodiversity	Net-positive ingredient footprint
Zero-Packaging Systems	On-site micro-dispensers, reusable vessels	Refill models for gyms/offices
Biodegradable Active Carriers	Biopolymer microcapsules	Fully compostable functional systems

Formulator's Insight:

"Zero waste" must extend to formulation design, every gram must add purpose or return value.

34.8 Sensory Evolution: Multisensory Beverage Experiences

The sensory landscape is expanding beyond taste and aroma to incorporate texture, color-changing effects, layered aromatics, and interactive sensory cues. These multisensory experiences appeal to emotion, novelty, and wellness signaling. Sensory evolution opens new pathways for consumer engagement and deeper brand differentiation.

Future beverages will appeal not just to taste, but to sight, sound, and emotion.

Table 34.8: Multisensory Beverage Experience Technologies

Innovation	Mechanism	Example
Dynamic Color Shift	pH-reactive botanicals (e.g., butterfly pea flower)	Visual freshness cue
Aroma-triggered caps	Aroma burst on opening	"First-sip" engagement
Temperature-responsive flavors	Encapsulated cooling or warming agents	Seasonal adaptation
Functional "sound branding"	Pairing specific sonic frequencies with flavor types	Digital beverage experiences

Neurosensory Insight:

Aroma and color pre-condition emotional expectation, "pink = calm," "yellow = alert."
The beverage of the future is a sensorial interface, not just a liquid.

34.9 Scientific and Regulatory Challenges Ahead

Emerging ingredients, novel processing technologies, and personalized nutrition platforms introduce new scientific and regulatory complexities. Ensuring safety, demonstrating efficacy, and aligning with evolving global frameworks will require deeper interdisciplinary collaboration. These challenges will shape how rapidly the next generation of beverages can reach the market.

Table 34.9: Scientific and Regulatory Challenges in Next-Gen Functional Beverages

Challenge	Impact	Needed Response
Evolving functional claims	Stricter evidence requirements (EU, US FDA)	Clinical substantiation and transparency
Novel ingredient regulation	New bioactives under review	GRAS / Novel Foods dossiers
Precision fermentation labeling	"Natural" claim ambiguity	Harmonized definitions
AI-driven formulation ethics	Data privacy, bias	Clear consumer consent protocols

Formulator's Insight:

Regulation will increasingly favor data-backed, safe innovation, not "claims chasing."
Future leaders will invest as much in validation science as in flavor development.

34.10 Vision: The Beverage as a Bio-Platform

The beverage of the future becomes a dynamic bio-platform, capable of delivering personalized nutrition, supporting cellular function, and integrating seamlessly with digital health ecosystems. This vision redefines beverages as adaptive wellness tools rather than static consumables, marking a transformative shift in their role within daily life.

Tomorrow's electrolyte drink will not be a static formula, it will be a living interface between biology, data, and sustainability.

The Future Beverage Ecosystem:

1. Digitally adaptive – AI monitors hydration, nutrient needs.
2. Bio-intelligent – Formulations co-designed with human physiology.
3. Regenerative – Ingredients that restore ecosystems.
4. Emotive and experiential – Flavor, color, and narrative harmonized.
5. Modular – Functional cores adapted per region, climate, or activity.

Formulator's Closing Insight:

The most advanced hydration system will be one that feels human, simple, transparent, adaptive, and symbiotic with nature.

34.11 Summary: From Function to Future

The evolution from basic hydration to highly engineered, multifunctional systems reflects the merging of biology, technology, and sustainability. The path forward emphasizes precision, personalization, and planetary responsibility. The future of functional beverages lies not only in delivering benefits, but in redefining what a beverage can be.

Table 34.11: Evolution of Functional Hydration: From Sports Drinks to Bioadaptive Wellness Systems

Past (2000–2015)	Present (2016–2025)	Future (2026–2035)
Sports drinks	Functional hydration	Bioadaptive wellness systems
Artificial colors/flavors	Natural and transparent	Regenerative and personalized
Static formulations	Sensory-optimized blends	AI-formulated modular systems
Shelf-stable	Fresh and smart	Connected and circular

Final Note:

Innovation in functional hydration is no longer about creating new flavors, it's about redefining what hydration means in an intelligent, regenerative, and emotionally satisfying world.

35. Epilogue - From Chemistry to Connection

The science of functional beverages began as chemistry: ions, pH, osmolality, and shelf stability.
But as this field matured, it revealed something more profound, hydration is not just a biochemical necessity; it's a human experience.

Every formulation in this book, every table and parameter, ultimately serves one purpose:
to deliver wellness in a form the body recognizes and the senses trust.

What started as the quest for efficiency, replenishing electrolytes, stabilizing vitamins, extending shelf life, has become a journey toward empathy in formulation.
Each ingredient, from the simplest salt to the most sophisticated botanical extract, participates in a quiet dialogue between biology and emotion.

The New Role of the Beverage Scientist

Tomorrow's beverage scientist is not merely a formulator.
They are:

- A translator of systems: connecting biochemical pathways with human perception.
- A sustainability architect: balancing product innovation with planetary boundaries.
- A sensory storyteller: crafting flavor experiences that evoke trust and delight.
- A data artisan: using analytics not to replace intuition, but to refine it.

As automation, AI, and bioengineering evolve, the competitive advantage will not be technology alone, but the human touch behind the formulation, curiosity, responsibility, and creativity.

From Function to Meaning

The drinks of the next decade will not compete on sodium levels or vitamin counts. They will compete on purpose, transparency, and resonance.

A successful beverage will:

1. Respect the body's intelligence: supporting hydration without overwhelming it.
2. Honor the planet's balance: through circular design and responsible sourcing.
3. Reflect the consumer's identity: clean, modern, authentic.

Functional hydration is becoming functional humanity, a medium where science serves wellness and design celebrates integrity.

The Closing Thought

If water is life, then functional hydration is the art of sustaining it.
From the molecular to the emotional, from the source to the sip, every decision a beverage developer makes, ingredient, process, or packaging, becomes part of a larger ecosystem of care.

As this field continues to evolve, let every new drink begin with one enduring question:
"Does this formula help someone live better, think clearer, and feel more connected?"

If the answer is yes, then it's not just a beverage.
It's progress, distilled.

36. Appendices

36.1 Appendix A: Glossary of Terms

Absorption Rate

The speed at which water, electrolytes, and nutrients pass from the gastrointestinal tract into systemic circulation. Influenced by osmolality, solute type (e.g., glucose–sodium co-transport), and gastric emptying.

Acid–Sweet–Salt Triangle

A sensory balancing framework where acidity, sweetness, and saltiness interact to create perceived refreshment, palatability, and functional authenticity in hydration beverages.

Adaptogen

A natural botanical compound (e.g., ashwagandha, rhodiola) that supports physiological homeostasis under physical or mental stress, typically by modulating cortisol and neurotransmitter activity.

Aqueous Extract

A botanical concentrate prepared by water-based extraction, favored in clean-label beverage systems for solvent-free, GRAS-compliant actives.

Aquaporin

Specialized protein channels in cell membranes (e.g., intestinal epithelium) that facilitate rapid transcellular water transport, often upregulated by hydration status and certain nutrients.

Bioavailability

The fraction of an ingested nutrient or electrolyte that is absorbed and becomes physiologically active. Enhanced by molecular form (e.g., citrate vs. sulfate salts), co-nutrients (e.g., vitamin C with iron), and delivery system (e.g., encapsulation).

Buffer System

A combination of weak acid and its conjugate base (e.g., citric acid + sodium citrate) that resists pH changes during processing and storage, preserving flavor, clarity, and microbial safety.

Carbohydrate Oxidation Rate

The metabolic rate at which ingested carbs are converted to energy (ATP), measurable via indirect calorimetry. Maximized (~1.0–1.6 g/min) using dual-transporter blends (e.g., glucose + fructose).

Chelation

Binding of metal ions (e.g., Ca^{2+}, Fe^{2+}, Cu^{2+}) by organic ligands (e.g., citrate, gluconate, EDTA) to prevent oxidation, haze formation, and off-flavors, critical for color and vitamin stability.

Clean-Label

A formulation and marketing approach emphasizing recognizable, naturally derived ingredients, minimal processing, and transparent declarations (e.g., "sweetened with stevia and fruit sugars").

Electrolyte Matrix

The intentional blend of ionic salts (Na^+, K^+, Mg^{2+}, Ca^{2+}, Cl^-) designed to match physiological needs (e.g., sweat loss), optimize absorption, and minimize sensory harshness.

Functional Osmolality Window

The target osmolality range (e.g., 270–320 mOsm/kg for isotonic drinks) that balances fluid absorption kinetics with energy delivery and gastric comfort.

Gastric Emptying Rate

The speed at which beverage contents leave the stomach and enter the small intestine, modulated by osmolality, volume, temperature, and macronutrient composition.

Hurdle Technology

A preservation strategy combining multiple mild inhibitory factors (e.g., low pH, natural antimicrobials, oxygen exclusion) to ensure microbial safety without harsh processing or synthetic additives.

Ionic Strength

A measure of total ion concentration in solution, influencing protein stability, flavor perception, and colloidal behavior. Expressed in mEq/L or mOsm/kg.

Isotonic

A beverage with osmolality matching that of human plasma (~280–320 mOsm/kg), enabling simultaneous hydration and energy delivery with minimal GI distress.

Hypotonic

A beverage with lower osmolality than plasma (<270 mOsm/kg), formulated for rapid fluid uptake, ideal for rehydration in heat, post-exercise, or clinical settings.

Hypertonic

A beverage with higher osmolality than plasma (>330 mOsm/kg), designed for post-exercise glycogen replenishment and nutrient density, typically consumed with additional water.

Masking (Flavor)

The use of complementary aromas, acids, sweeteners, or texturants to reduce or redirect perception of off-notes (e.g., bitterness from magnesium, sulfur from amino acids).

Metabolic Hydration

A formulation paradigm where fluid, electrolytes, and nutrients (e.g., carbs, amino acids) are co-optimized to support not just volume restoration, but cellular energy production and thermoregulation.

Microencapsulation

A delivery technique entrapping actives (flavors, vitamins, botanicals) within protective matrices (e.g., maltodextrin, gum arabic) to improve stability, control release, and mask sensory defects.

Osmoregulation

The body's continuous process of maintaining fluid and electrolyte homeostasis via hormonal signals (e.g., vasopressin), renal function, and thirst.

Osmolality

The concentration of osmotically active particles (ions, sugars) per kilogram of solvent, a key predictor of fluid absorption speed. Measured in milliosmoles per kilogram (mOsm/kg).

SGLT1 (Sodium-Glucose Cotransporter 1)

An intestinal transporter that couples sodium and glucose (or galactose) uptake, driving rapid water absorption, leveraged in oral rehydration and performance beverages.

Sensory Flow Curve

The dynamic evolution of mouthfeel perception during sipping, initial contact (viscosity), mid-palate (acidity/body), and aftertaste (cleanness), engineered via carbohydrate and hydrocolloid selection.

Shelf-Life Engineering

The integrated design of formulation, processing, and packaging to ensure consistent safety, stability, and sensory quality over the product's intended lifetime, validated via real-time and accelerated studies.

Sodium–Potassium Ratio (Na^+:K^+)

A critical electrolyte design parameter. A 2:1 to 3:1 ratio supports extracellular rehydration and thirst drive; lower ratios (e.g., 1.5:1) favor intracellular recovery post-exercise.

Sweetness Architecture

The strategic layering of sweeteners (e.g., fast-onset fructose, mid-body allulose, long-tail stevia) to create balanced, natural-tasting sweetness with minimal caloric or osmotic load.

Thermal Degradation

Breakdown of sensitive compounds (vitamins, flavors, polyphenols) due to heat exposure; mitigated by process optimization (e.g., HTST vs. batch pasteurization) and protective additives.

Turbidity (NTU)

A measure of optical clarity in liquids, expressed in Nephelometric Turbidity Units. Target: <1.5 NTU for clear functional beverages; higher values acceptable in recovery or plant-based formats.

Water Activity (a_w)

The ratio of vapor pressure of water in a product to that of pure water; influences microbial growth and chemical stability. Reduced by solutes (sugars, salts).

36.2 Appendix B: Electrolyte Reference Tables

Table B.1: Physiological Electrolyte Loss in Sweat (Typical Ranges)

Values represent mg/L of sweat; individual variation may be ±30%.

Ion	Low Sweat (mg/L)	Moderate Sweat (mg/L)	High Sweat (mg/L)	Primary Physiological Role
Na^+	400–600	800–1,200	1,200–1,800	Plasma volume maintenance, nerve transmission, osmotic drive for water absorption via SGLT1
K^+	100–150	150–250	250–400	Intracellular fluid balance, muscle repolarization, glycogen synthesis
Cl^-	800–1,000	1,000–1,500	1,500–2,000	Charge balance, gastric acid synthesis, osmotic regulation
Mg^{2+}	5–15	10–30	20–50	ATP stabilization, muscle relaxation, enzyme cofactor (e.g., glycolysis)

Ca^{2+}	10–30	20–60	40–80	Muscle contraction signaling, enzyme activation, bone metabolism

Note: "Salty sweaters" may lose >2,000 mg Na^+/L. Sweat Na^+ correlates with dietary Na^+ intake, acclimatization, and genetics.

Table B.2: Electrolyte Salt Comparison for Beverage Formulation

Ranking of common salts by solubility, taste, stability, and functional suitability.

Salt (Anhydrous)	**Ion Delivered**	**Solubility (g/100 mL, 25 °C)**	**Taste Profile**	**pH Effect**	**Bioavailability**	**Best Use Context**
Sodium chloride	Na^+, Cl^-	36.0	Strong salty	Neutral	★★★★★	High-sweat rehydration, performance base
Sodium citrate	Na^+	73.0*	Mild salty–tart	↑ pH	★★★★☆	Isotonic/hypotonic; buffers, masks harshness
Potassium chloride	K^+, Cl^-	34.0	Bitter, metallic	Neutral	★★★☆☆	Use <0.3 g/L; blend with citrate
Potassium citrate	K^+	57.5*	Soft, tart	↑ pH	★★★★☆	Preferred for taste & buffering

Magnesium chloride	Mg^{2+}, Cl^-	54.3	Bitter, metallic	Neutral	★★★☆☆	≤0.05% w/v; pair with flavor masking
Magnesium lactate	Mg^{2+}	~10 (pH 4.0)	Mild, smooth	Slight ↑	★★★★☆	Premium isotonic/recovery; low bitterness
Calcium lactate	Ca^{2+}	9.0	Neutral	Slight ↑	★★★★☆	Recovery drinks; compatible with fruit acids
Calcium gluconate	Ca^{2+}	3.5	Very mild	Neutral	★★★★☆	Clear beverages; low turbidity risk
Trisodium citrate	Na^+	73.0*	Salty–tart	↑↑ pH	★★★★☆	Clinical ORS, buffering systems

* Highly soluble but hygroscopic; pre-dissolve in warm water.
Formulator Tip: Citrate/lactate salts enhance clarity and reduce precipitation risk in acidic beverages (pH 3.5–4.0). Chlorides provide faster plasma expansion but require robust flavor masking.

Table B.3: Target Electrolyte Ranges by Beverage Function

All values in mg/L of finished beverage (elemental ion basis).

Function & Type	Na^+	K^+	Mg^{2+}	Ca^{2+}	Cl^-	Target Osmolality (mOsm/kg)	Notes

Hypotonic Rehydration	150–300	80–150	5–15	5–15	200–400	150–250	Fast gastric emptying; use Na-citrate/K-citrate
Isotonic Performance	250–400	120–200	10–25	10–25	300–600	270–310	Balance with 4–8% carbs; $Na^{+}:K^{+} \approx$ 2.5:1
Hypertonic Recovery	350–550	150–300	15–40	15–40	500–900	330–380	Higher Na supports fluid retention with carbs/protein
Clinical ORS (WHO-type)	~1,700*	~780*	50–100*	-	~2,000*	~245	*Per WHO rehydration solution (mmol-based); not for sports use
Heat-Exposure/Heavy Sweat	400–750	150–300	20–40	20–40	700–1,200	300–320	Add citrate buffer; stronger flavor needed to offset salinity

Caution: Exceeding 500 mg Na^+/L in low-carb beverages increases perceived saltiness; adjust acidity (↑ citric/malic) and sweetness (↑ clean sweeteners) accordingly.

Table B.4: Electrolyte–Acid–Flavor Compatibility Matrix

Guidance for minimizing off-notes and maximizing stability.

Electrolyte	**High-Risk Interactions**	**Preferred Acid System**	**Recommended Flavor Families**	**Stabilization Tips**
Na^+ (from NaCl)	Bitterness amplification with stevia	Citric + malic (2:1)	Citrus, berry, tropical	Use Na-citrate blend (≥50%); keep pH 3.6–3.8
K^+ (from KCl)	Metallic, lingering bitterness	Malic > citric	Mango, pomegranate, hibiscus	Mask with 0.05% vanilla or citrus top-note oil
Mg^{2+}	Astringency, chalkiness	Lactic or malic	Pineapple, coconut, mint	Use Mg-lactate; pair with glycerol (0.2%) for smoothness
Ca^{2+}	Precipitation with phosphates/sulfates	Citric (buffered)	Vanilla, banana, cocoa	Avoid pH <3.5; use lactate/gluconate forms; limit ≤30 mg/L in clear drinks
Cl^-	"Briny" mouthfeel at high load	Citrate buffer	Lemon-lime, cucumber	Combine with Na-citrate to soften perception

Stability Note: Ca^{2+} and Mg^{2+} >50 ppm total in acidic drinks (pH <3.8) increase haze risk. Use chelators (sodium citrate 0.1–0.2%) and filter to 0.45 μm.

Table B.5: Electrolyte Stability & Precipitation Risk Matrix

Common failure modes and mitigation strategies.

Risk	Typical Cause	Critical Threshold	Prevention Strategy
Mineral haze/sediment	Ca/Mg + phosphate/sulfate at low pH	Ca^{2+} >40 ppm + PO_4^{3-} >20 ppm @ pH <4.0	Use lactate/gluconate salts; avoid trisodium phosphate in Ca/Mg systems
Crystallization in concentrates	Supersaturation upon cooling	TDS >5% w/v, [Na^+] >800 mg/L	Dilute base; use blended salts (citrate reduces crystallization tendency)
Metal-catalyzed oxidation	Trace Fe/Cu in salts reacting with ascorbate/polyphenols	Fe >0.1 ppm	Use purified salts; add citrate (0.1%) or EDTA (≤30 ppm)
Protein–electrolyte flocculation	Whey near isoelectric point (pH 5.2) + Ca^{2+}	pH 4.8–5.5 + Ca >20 ppm	Maintain pH 3.8–4.0 for clear whey; use hydrolyzed collagen for acid stability
Bitterness drift on storage	Hydrolysis of $MgCl_2$/KCl over time	>6 months at 30 °C	Replace ≥70% chloride salts with organic acid salts

Table B.6: Elemental Ion Contribution Factors for Common Electrolyte Salt

Conversion Table: Salt → Elemental Ion

For accurate dosing. Multiply salt weight by factor to get elemental ion (mg).

Salt (Anhydrous)	Na^+	K^+	Mg^{2+}	Ca^{2+}	Cl^-
Sodium chloride	0.393	-	-	-	0.607
Sodium citrate	0.164	-	-	-	-
Potassium chloride	-	0.524	-	-	0.476
Potassium citrate	-	0.358	-	-	-
Magnesium chloride	-	-	0.257	-	0.743
Magnesium lactate	-	-	0.122	-	-
Calcium lactate	-	-	-	0.130	-
Calcium gluconate	-	-	-	0.089	-

Example: To deliver 200 mg Na^+ using sodium citrate:
200 mg $Na^+ \div 0.164 = 1{,}220$ mg sodium citrate

Table B.7: Electrolyte Compatibility Guide by Beverage Format

Format	Recommended Electrolyte Strategy	Critical Constraints
RTD Bottles (PET)	Na-citrate + K-citrate + Mg-lactate	Avoid high Cl^- if carbonated; use UV barrier for natural colors
Powders	Spray-dried NaCl/KCl blends; microencapsulate Mg/Ca	Control hygroscopicity (use maltodextrin carrier; RH <40%)
Effervescent Tablets	$NaHCO_3$ + K-citrate + Mg-oxide	Limit acid load (citric/tartaric); avoid Ca in high-acid tablets

Concentrates (10×)	Chelated forms (Na-gluconate, Mg-malate)	Keep TDS <15%; refrigerate post-dilution
Carbonated Hydration	Low-ionic strength (<300 mg/L Na^+)	High osmolality reduces CO_2 retention; avoid Ca/Mg >10 ppm

36.3 Appendix C - Ingredient Function and Compatibility Chart

Key to Symbols:

✅ = Highly compatible / low risk

⚠ = Use with caution / moderate interaction risk

✖ = Avoid combination or high incompatibility

◦ = pH- or concentration-dependent

• = Requires encapsulation, chelation, or process control

Table C.1: Ingredient Function and Compatibility

Ingredient	Primary Function	Solubility (in Water)	Optimal pH Range	Thermal Stability	Key Compatibility Notes
Sodium Chloride (NaCl)	Plasma volume restoration, SGLT1 co-transport driver	Very high (~36 g/100 mL)	3.0–7.0	High	⚠ Bitterness/saltiness at >0.3% w/v; ✖ avoid with high Mg^{2+}/Ca^{2+} unless buffered;

					✅ pairs well with citric acid & citrus flavors
Sodium Citrate	Buffer, Na^+ source, reduces salt harshness	Moderate (~9 g/100 mL)	3.5–6.5	High	✅ Excellent with K^+, Mg^{2+}, Ca^{2+}; stabilizes anthocyanins; ◦ watch pH rise in low-acid systems
Potassium Chloride (KCl)	Intracellular balance, muscle function	High (~34 g/100 mL)	3.0–7.0	High	✖ Bitter/metallic beyond 0.15% w/v; ⚠ mask with malic acid or berry flavors; ✅ blend with K-citrate
Potassium Citrate	Smooth K^+ delivery, lactate buffer	Moderate (~20 g/100 mL)	3.5–5.5	High	✅ Ideal for clear isotonic drinks; synergistic with citric/malic acids; improves Mg^{2+} solubility
Magnesium Chloride	ATP stabilization, cramp prevention	High (~54 g/100 mL)	3.0–7.0	High	✖ Bitter/metallic

					even at 0.05% w/v; ⚠ avoid in clear beverages unless blended with Mg-lactate; ✅ use in opaque recovery
Magnesium Lactate	Bioavailable Mg^{2+}, smooth mouthfeel	Moderate (~10 g/100 mL)	3.5–5.0	Moderate	✅ Preferred for clear beverages; low bitterness; ⚠ avoid pH <3.5 (haze risk); compatible with most flavors
Calcium Lactate	Muscle signaling, bone support	Moderate (~9 g/100 mL)	4.0–6.5	High	✅ Stable in fruit acids; ⚠ may cause slight turbidity; ✖ avoid with phosphates or high sulfate
Calcium Gluconate	Low-taste Ca^{2+}, clean-label option	Moderate (~3.5 g/100 mL)	3.5–6.0	High	✅ Best for clear systems; pre-dissolve in warm water; ◦

					limit to <40 ppm for clarity
Citric Acid	Acidity, flavor brightness, chelation	Very high	2.5–6.0	High	✅ Universal acidulant; enhances sweetness & Na^+ absorption; ⚠ may invert sucrose >85 °C
Malic Acid	Smooth tartness, fruit authenticity	High	3.0–5.5	High	✅ Ideal for apple/berry; softens K^+/Mg^{2+} bitterness; synergistic with citrate buffers
Lactic Acid	Mild tang, recovery support	High	3.5–5.0	High	✅ Works well in protein/recovery drinks; adds body; less volatile than citric/malic
Glucose (Dextrose)	Rapid energy, osmotic hydration	High	3.0–7.0	Moderate	✅ Drives SGLT1; ⚠ high osmolality; ✗ degrades vitamin C if unchelated

Maltodextrin (DE 10–20)	Sustained energy, mouthfeel	High	3.0–7.0	High	✅ Low sweetness, isotonic balancer; ✅ stabilizes proteins & aromas; ❌ may hydrolyze at pH <3.2, high heat
Isomaltulose	Slow-release carb, low glycemic	High	3.5–6.0	Moderate	✅ Clean-label performance carb; ⚠ Maillard browning with proteins; avoid >85 °C >1 min
Stevia (Reb A/M/D)	Zero-cal sweetener	Low (extract); soluble forms available	3.0–5.0	Moderate (Reb M > Reb A)	⚠ Bitter aftertaste alone; ✅ blend with erythritol or citrus acids; ❌ avoid high NaCl without masking
Erythritol	Cooling sweetness,	High	3.0–6.0	High	✅ Enhances freshness;

	bulking agent				reduces bitterness; ⚠ crystallization risk at low temps/high concentration
Allulose	Sugar-like mouthfeel, low-cal	High	3.0–6.0	Moderate	✅ Improves flavor roundness; ⚠ caramelizes >90 °C; ✅ pairs well with stevia
Whey Protein Isolate	Muscle repair, satiety	Excellent (pH 3.0–4.0 or >6.0)	3.5–4.0 (acidic)	Moderate	✅ Stable in acidic RTD; ⚠ denatures near pI (5.2); ❌ avoid high Ca^{2+}/Mg^{2+} without citrate buffer
Collagen Peptides	Skin/joint support, clean protein	Excellent (pH 3–7)	3.5–4.5	High	✅ Gold standard for clear functional waters; ✅ compatible with

					Na^+/K^+/vitamin C; no haze risk
Gellan Gum (Low-Acyl)	Suspension, clarity, low-viscosity texture	High (needs heat hydration)	3.0–7.0	High	✅ Ideal for clear hydration; stabilizes minerals; ✅ synergistic with citrate buffers
Xanthan Gum	Viscosity, emulsion stability	High	2.5–8.0	High	⚠ Can cause slimy texture >0.05%; ❌ avoid in sparkling systems (foam stability); ✅ fine for recovery
Gum Arabic	Flavor emulsifier, mouthfeel smoother	High	3.0–6.5	High	✅ Excellent for clear citrus/aromatic systems; stabilizes volatile oils & polyphenols
Ascorbic Acid (Vit C)	Antioxidant, freshness cue	Very high	<4.5	Low (oxidizes with	⚠ Degrades rapidly if unchelated;

				heat/O_2/metals)	✅ use with Na-citrate/EDTA; ❌ avoid with Fe/Cu traces or high pH
Green Tea Extract (EGCG)	Antioxidant, metabolism	Moderate (aqueous)	3.5–4.2	Low	⚠️ Bitter/astringent; oxidizes fast; ✅ stabilize with ascorbate + citrate; ❌ avoid high Ca^{2+}/heat without encapsulation
Ashwagandha (Aqueous)	Adaptogen, stress support	Moderate	4.0–6.0	Low	❌ Strong earthy/bitter note; ⚠️ use ≤0.03% + citrus/vanilla masking; ✅ prefer encapsulated forms
L-Theanine	Calm focus, caffeine modulator	Excellent	3.0–7.0	High	✅ Highly compatible; no stability issues; synergistic with

					citrus/tea flavors and low-dose caffeine
Caffeine	Alertness, endurance	High	3.0–7.0	High	✅ Stable in most matrices; ⚠️ bitterness at >100 mg/L; ✅ pair with theanine (2:1 ratio) for clean energy
Anthocyanins (e.g., hibiscus)	Color + antioxidant	Moderate	2.5–4.0	Moderate	❌ Color shifts >pH 4.2; ⚠️ light/heat-sensitive; ✅ stabilize with ascorbate, citrate, amber packaging
β-Carotene	Natural yellow/orange color	Insoluble (requires emulsion)	3.0–7.0	High (in emulsion)	✅ Use gum arabic/polysorbate emulsions; ❌ avoid in clear systems without nanoemulsification
Potassium Sorbate	Mold/yeast inhibitor	High	<5.5	Moderate	✅ Clean-label preservative;

synergistic with low pH & citrate;

⚠ slight odor at >0.08%

High-Risk Incompatibility Summary (Avoid or Mitigate):

Table C.2: High-Risk Incompatibility Summary

Risk Combination	Observed Issue	Mitigation Strategy
Ca^{2+} + Phosphate	White precipitate, chalkiness	❌ Replace phosphate with citrate/lactate; use Ca-gluconate; pH >4.0
Mg^{2+} + High Cl^{-}/SO_4^{2-}	Metallic bitterness, haze	⚠ Limit Mg-Cl to ≤0.05%; use Mg-lactate/gluconate; add citrate buffer
Protein + Ca^{2+} at pH ~4.0	Aggregation/sediment	✅ Use Na-citrate chelation; keep Ca <50 ppm; pre-hydrate protein before acidifying
Polyphenols + Fe/Cu traces	Rapid browning, off-flavor	✅ Chelate with citric acid/EDTA; nitrogen flush; avoid stainless steel contact
Stevia + High NaCl	Amplified bitterness	✅ Add citric acid + vanillin/erythritol; use Reb M > Reb A
Fructose + Whey Protein	Maillard browning (ambering)	⚠ Use <3% fructose; cold-fill; avoid >85 °C; add tocopherol

Vitamin C + Copper/Magnesium	Rapid oxidation	✅ Use purified salts (<0.1 ppm metals); add chelators; package in O_2-barrier

Formulator's Quick-Reference Guide:

Table C3: Formulator's Quick-Reference

Beverage Type	Recommended Core Matrix	Avoid
Hypotonic Rehydration	Na-citrate + K-citrate + Mg-lactate + citric + low glucose + gum arabic	Chlorides >0.1%, high viscosity agents, phosphates
Isotonic Performance	NaCl + Na-citrate + K-citrate + Mg-lactate + glucose/maltodextrin + citric/malic	Unbuffered acids, high fructose, unchelated minerals
Recovery (Hypertonic)	Na-citrate + K-citrate + Ca-lactate + maltodextrin + whey/collagen + lactic acid	Ca-phosphate, high heat without stabilization, unmasked BCAAs
Clean-Label Functional Water	Sea salt extract + K-citrate + Mg-lactate + isomaltulose/erythritol + gum arabic + botanical extracts (encapsulated)	Synthetic preservatives, artificial colors, high TDS (>1.5%)
Botanical Antioxidant Drink	Citrate buffer + ascorbate + anthocyanins + green tea (low dose) + erythritol + gum arabic	High Ca^{2+}/Fe, clear PET (use amber), hot-fill without antioxidants

36.4 Appendix D - Example Formulation Templates

Table D1: Basic Electrolyte Rehydration Drink (Hypotonic)

Purpose: Rapid fluid absorption for light activity, heat recovery, or daily wellness.
Target: 200–230 mOsm/kg, pH 3.7, <10 kcal/100 mL, shelf life: 12 months.

Ingredient	% w/w	Function / Rationale
Purified water (deaerated)	q.s. to 100	Base fluid
Sodium chloride	0.045	Replaces sweat Na^+ (~180 mg/L)
Potassium citrate	0.020	Muscle support, smooth K^+ delivery
Glucose (anhydrous)	2.00	Enables SGLT1 co-transport, mild energy
Citric acid	0.10	pH control, freshness, chelation
Natural lemon–lime flavor	0.10	Palatability, top-note aroma
Stevia (Reb M, 97%)	0.015	Zero-cal sweetness boost
Optional: Ascorbic acid	0.02	Antioxidant, freshness cue
Optional: Gum Arabic	0.05	Stabilizer for clarity & mouthfeel

Processing & Stability Notes

- Mix salts first in water → add glucose → acidify to pH 3.7 ± 0.05 → cool → add flavors/sweeteners.
- HTST: 82 °C, 25 s; cool to ≤25 °C in ≤2 min.
- Fill under N_2 flush (DO < 0.5 mg/L).
- Packaging: Amber PET (EVOH barrier) or aluminum can.
- Validation: Turbidity <1.0 NTU, ΔE <1.5 (6 mo), pH drift <0.05.

Table D.2: Classic Sports Isotonic Drink

Purpose: Balanced hydration + energy for endurance or team sports.
Target: 290 mOsm/kg, pH 3.6–3.8, 26 kcal/100 mL, 12-month ambient shelf life.

Ingredient	% w/w	Function / Rationale
Purified water	q.s. to 100	Solvent
Maltodextrin (DE 15)	4.00	Sustained energy, low sweetness
Glucose	2.00	Rapid fuel, osmotic driver
Sodium chloride	0.090	Primary Na^+ source (~360 mg/L)
Potassium citrate	0.035	Intracellular balance
Citric acid	0.20	Acidity, flavor brightness
Natural orange flavor	0.10	Consumer acceptability
Beta-carotene (emulsion 10%)	0.005	Natural yellow hue
Ascorbic acid	0.03	Antioxidant, label appeal
Potassium sorbate	0.03	Natural preservative (≤0.05% compliant)

Processing & Stability Notes

- Hydrate maltodextrin in warm water (45 °C) → add salts → acidify → cool → add flavor/color.
- Homogenize 2× at 100 bar for uniform dispersion.
- Hot-fill: 88 °C, hold 15 s → cool immediately.
- Stability: Osmolality 285–295 mOsm/kg post-fill; no haze after 40 °C/3 mo test.

Table D.3: Caffeine-Enhanced Energy Hydration (Isotonic Hybrid)

Purpose: Mental + physical performance for esports, precision sports, or afternoon fatigue.
Target: 300 mOsm/kg, pH 3.7, 18 kcal/100 mL, calm-alert profile.

Ingredient	% w/w	Function / Rationale
Water (deaerated)	q.s. to 100	Base
Glucose	2.0	Energy + co-transport
Isomaltulose	2.0	Slow-release carb, low glycemic
Sodium citrate	0.25	Buffer, smooth Na^+
Potassium chloride	0.10	Electrolyte balance
Caffeine (green tea sourced)	50 mg/L (0.005%)	Alertness, ≤200 mg/day daily dose
L-Theanine	100 mg/L (0.010%)	Calm focus, 2:1 ratio with caffeine
Citric acid	0.15	Brightness, pH 3.7
Natural citrus–mint flavor	0.03	Aroma lift, cooling perception
Stevia (Reb M) + erythritol	0.02 + 0.40	Natural sweetness, mouthfeel
Gum Arabic	0.05	Stabilizer, flavor fixative
Ascorbic acid	0.03	Antioxidant protection

Processing & Stability Notes

- Add caffeine/theanine after cooling to ≤35 °C.
- Avoid >85 °C as it degrades L-theanine over time.
- Cold-fill or aseptic preferred for amino stability.
- Shelf life: 9–12 months; validate sensory masking (bitterness <2/9 panel score).

Table D.4: Plant-Based Recovery Beverage

Purpose: Post-workout glycogen + protein replenishment; vegan, clean-label.
Target: 340 mOsm/kg, pH 4.0, 38 kcal/100 mL, smooth mouthfeel.

Ingredient	% w/w	Function / Rationale
Coconut water (single-strength)	80.0	Natural K^+/Mg^{2+}, light sweetness
Water	18.79	Dilution medium
Oat protein hydrolysate	1.0	Plant protein, low allergen
Cane sugar	3.0	Energy, mouthfeel
Sea salt	0.08	Na^+ source
Potassium citrate	0.03	Muscle recovery
Magnesium lactate	0.05	Cramp prevention
Citric acid	0.10	pH 4.0 adjustment
Natural cacao extract	0.02	Flavor, antioxidant
Xanthan gum	0.03	Mouthfeel, suspension
Optional: Vitamin D3 + B12 (encapsulated)	Trace	Recovery support

Processing & Stability Notes

- Pre-hydrate oat protein at 45 °C, 30 min, before acid addition.
- Use low-shear mixing post-pH adjustment to avoid foam.
- HTST: 85 °C, 20 s → homogenize 120 bar.
- Clarify to ≤2 NTU (slight opacity acceptable for recovery segment).
- Sensory target: creamy, lightly sweet, no beany off-notes.

Table D.5: Herbal Hydration Infusion (Clean-Label Functional Water)

Purpose: Daily wellness hydration with adaptogenic support; premium positioning. Target: 230 mOsm/kg, pH 3.8, <5 kcal/100 mL, transparent.

Ingredient	% w/w	Function / Rationale
Purified water	q.s. to 100	Base
Electrolyte blend (Na^+ 300 mg/L, K^+ 100 mg/L)	~0.15	Sea salt extract + K-citrate
Ashwagandha aqueous extract(5% withanolides)	150 mg/L (0.015%)	Stress resilience
Green tea extract(40% EGCG)	100 mg/L (0.010%)	Antioxidant, mild caffeine
Honey solids	1.0	Natural sweetness, prebiotic
Citric acid	0.08	pH 3.8, brightness
Natural lemon balm + ginger flavor	0.03	Aroma, adaptogenic synergy
Glycerol	0.25	Mouthfeel, bitterness masking
Gum Arabic	0.05	Stabilizer
Ascorbic acid	0.03	Protects polyphenols

Processing & Stability Notes

- Add botanicals post-cooling to ≤30 °C.
- Encapsulate ashwagandha (maltodextrin carrier) for clarity.
- Cold-fill with N_2 headspace.
- Packaging: Amber glass or UV-PET.
- Shelf life: 9 months (monitor polyphenol retention ≥85%).

Table D.6: Electrolyte Sparkling Water (Zero-Cal, RTD)

Purpose: Light hydration with effervescence; lifestyle/convenience format.
Target: 220 mOsm/kg, pH 3.6, 0 kcal, high clarity.

Ingredient	% w/w	Function / Rationale
Carbonated water (CO_2 3.0–3.5 vol)	q.s. to 100	Base + sensory lift
Sodium bicarbonate	0.04	Alkaline Na^+ source, CO_2 stabilizer
Potassium chloride	0.025	Electrolyte balance
Citric acid	0.08	pH 3.6, tartness
Natural fruit extract (lime, cucumber)	0.10	Flavor identity
Stevia (Reb M) + monk fruit	0.012 + 0.003	Clean sweetness
Optional: Sucralose (trace)	0.005	Flavor synergy
Ascorbic acid	0.02	Antioxidant, freshness

Processing & Stability Notes

- Carbonate after full chilling (≤4 °C).
- Minimize shear: avoid homogenization (foam risk).
- Cap torque: 1.8 ± 0.2 N·m for CO_2 retention.
- Test CO_2 loss: ≤10% after 3 months at 25 °C.
- Clarity: <0.5 NTU; no oil ring formation.

Table D.7: BCAA-Enriched Endurance Recovery Drink (Clear RTD)

Purpose: Muscle protection, reduced soreness after prolonged exertion.
Target: 360 mOsm/kg, pH 3.8, 30 kcal/100 mL, minimal bitterness.

Ingredient	% w/w	Function / Rationale
Water	q.s. to 100	Solvent
Maltodextrin	6.0	Glycogen replenishment
BCAA (Leu:Ile:Val = 2:1:1)	1.0% (10 g/L)	MPS activation
Sodium chloride	0.10	Hydration support
Potassium citrate	0.04	Cellular recovery
Magnesium lactate	0.01	Metabolism cofactor
Citric acid + malic acid (2:1)	0.20	Mask bitterness, round acidity
Natural tropical flavor (mango–passionfruit)	0.05	Aroma diversion
Erythritol + stevia	0.30 + 0.015	Sweetness without calories
Glycerol	0.25	Mouthfeel smoothing
Gum Arabic	0.05	Stabilization

Processing & Stability Notes

- Pre-dissolve BCAA at 40 °C under agitation.
- Add acids after BCAA dissolution to minimize foaming.
- Cold-fill or short HTST (80 °C, 15 s).
- Triangle test vs. control: <30% panelists detect bitterness.

Table D.8: Vitamin–Mineral Enhanced Hydration Drink (Daily Wellness)

Purpose: Nutrient-fortified hydration for office, travel, or elderly use.
Target: 240 mOsm/kg, pH 3.8, 12 kcal/100 mL, stable vitamin retention.

Ingredient	% w/w	Function / Rationale
Water	q.s. to 100	Base
Sodium chloride	0.06	Hydration anchor
Potassium citrate	0.025	Muscle/nerve support
Magnesium sulfate	0.005	Metabolism (use lactate if bitterness observed)
Vitamin C (ascorbic acid)	80 mg/L (0.008%)	Antioxidant, immune
Niacin (B3)	5 mg/L	Energy metabolism
Pyridoxine HCl (B6)	0.5 mg/L	Neurotransmitter support
Cyanocobalamin (B12, encapsulated)	1.2 µg/L	Cellular energy
Citric acid	0.10	pH 3.8
Erythritol	0.40	Low-cal sweetness
Stevia (Reb M)	0.02	Boost
Natural berry flavor	0.03	Consumer appeal
Gum Arabic	0.05	Stabilizer

Processing & Stability Notes

- Add B12 and ascorbic acid last, post-cooling.
- Use multilayer PET with O_2 barrier + N_2 flush.
- Validate: ≥90% C & B-vitamin retention at 6 months (HPLC).
- Label claim example: “Supports energy metabolism and daily hydration.”
- Formulator’s Quick Reference Guide

Table D.9: Formulator's Quick Reference Guide: Template Selection, Key Innovation, and Shelf-Life Strategy

Template	Best For	Key Innovation	Shelf-Life Tip
D.1 Hypotonic Rehydration	Heat stress, yoga, travel	Minimalist, fast absorption	N_2 flush + citrate buffer
D.2 Classic Isotonic	Mass-market sports	Dual-carb energy + clarity	Homogenize before hot-fill
D.3 Caffeine–Theanine	Focus drinks, esports	Cognitive hydration synergy	Cold-fill to protect amino acids
D.4 Plant Recovery	Vegan, sustainability	Coconut + oat protein matrix	Pre-hydrate protein to avoid grit
D.5 Herbal Infusion	Premium wellness	Adaptogen + polyphenol blend	Amber glass for light protection
D.6 Sparkling Water	Lifestyle, RTD growth	Zero-cal effervescence	Control CO_2 pressure + cap torque
D.7 BCAA Recovery	Endurance athletes	Clear amino delivery	Use glycerol + tropical flavor masking
D.8 Vitamin Hydration	Daily functional use	Multi-vitamin stability	Encapsulate B12, add ascorbate synergy

36.5 Appendix E: pH, Osmolality, and Stability Reference Data

E1: pH Ranges

Table E1.1: Preferred pH Ranges by Beverage Type

Beverage Type	Typical pH Range	Rationale
Isotonic hydration drinks	3.5 – 4.0	Microbial safety, flavor brightness
Electrolyte waters (low sweetness)	3.3 – 4.2	Mineral solubility + clarity
Collagen beauty waters	3.6 – 3.9	Prevent peptide aggregation
Plant or botanical tonics	3.2 – 4.0	Polyphenol stability, color control
Protein recovery drinks	3.4 – 3.8	Avoid protein coagulation
Electrolyte shots	3.2 – 3.6	High stability in concentrated systems
Vitamin waters	3.4 – 3.9	Ascorbic acid protection
Natural juice–electrolyte blends	3.1 – 4.0	Yeast/mold control without sorbate overload

Table E1.2: pH–Ingredient Compatibility Matrix

Ingredient Type	Stability at Low pH (3.2–3.8)	Notes
Sodium citrate	Excellent	Buffering capacity improves stability
Potassium chloride	Excellent	No hydrolysis or precipitation
Magnesium lactate	Good	Slight flavor impact at pH <3.4
BCAAs	Moderate	Lower pH improves solubility but increases aftertaste
Collagen peptides	Good	Optimum clarity at pH 3.6–3.8
Stevia glycosides	Excellent	No degradation at low pH

Anthocyanin colors	pH-sensitive	Red at pH <3.5; fades above 4.0
Turmeric/curcumin	Poor	Unstable below pH 3.5
Citrus flavors	Excellent	Acid enhances aroma volatility
Botanical extracts	Variable	Tannins may precipitate below pH 3.0

Table E1.3: pH Troubleshooting Guide

Symptom	**Likely Cause**	**Solution**
Precipitation after 7–14 days	Magnesium/calcium salts instability	Raise pH to >3.5; switch to gluconates or lactates
Color fading	Oxidation or pH >4	Improve oxygen control; increase acidity
Loss of aroma	pH drifting upward	Add stronger buffer system
Increased bitterness	pH <3.4	Increase citrate; rebalance acids
Phase separation	Protein/mineral interaction	Adjust pH to the safe zone (3.6–3.8)

E2. Osmolality Reference Data

Table E2.1: Osmolality Ranges by Beverage Category

Beverage Type	**Target Osmolality (mOsm/kg)**	**Interpretation**
Hypotonic hydration water	100 – 200	Faster absorption; ideal for heat conditions
Isotonic sports drink	260 – 320	Matches body fluid; optimized for exercise

Light functional water	180 – 260	Electrolyte boost without heaviness
Recovery drink (carb + protein)	300 – 400	Higher solids → higher osmolality
Electrolyte concentrates/shots	450 – 650	High-density actives; slow gastric emptying
Kids hydration drinks	200 – 260	Gentler on stomach; avoids overload

E2.2: Contribution of Electrolytes to Osmolality

Table E2.2: Approximate osmolality contributions (per liter):

Ingredient	Typical Level	Osmolality Contribution
Sodium (Na^+) 50–150 mg	+10–25 mOsm	High ionic activity
Potassium (K^+) 100–200 mg	+8–15 mOsm	Moderate
Magnesium (Mg^{2+}) 10–50 mg	+5–12 mOsm	Increases taste complexity
Sugars (3–5% sucrose)	+120–200 mOsm	Major driver of isotonicity
Polyols (erythritol)	+40–80 mOsm	Depends on dosage
BCAAs 1–3%	+20–60 mOsm	Adds bitterness if too high
Collagen peptides 0.5–1%	+10–20 mOsm	Minor effect

Quick Calculation Rule:

Every 1% sugar contributes approximately 35–40 mOsm/kg.

E2.3 Osmolality Troubleshooting Guide

Table E2.3: Osmolality Troubleshooting Guide

Symptom	Potential Cause	Adjustment

Drink feels "heavy"	Osmolality >330	Reduce sugar or polyols
Delayed absorption	Too many dissolved solids	Target hypotonic 150–250 mOsm
Harsh mineral taste	High ionic load	Reduce KCl; use citrate salts
Phase separation in cold	High protein or polyols	Lower concentration or add stabilizer
Flavor dullness	Too low osmolality (<150)	Increase organic acids or aroma

E3.: Stability Reference Tables

E3.1: Stability Thresholds for Key Ingredients

Table E3.1: Stability Thresholds for Key Ingredients

Ingredient	Stability Window	Notes
Ascorbic acid	Best at pH 3.0–3.6; low DO	Highly oxidation-sensitive
B vitamins	Stable at pH 3.5–4	Sensitive to light and heat
Electrolyte salts	Stable across pH 2–6	Taste shifts at extremes
Natural colors (anthocyanin)	Best < pH 3.3	Heat/light sensitive
Curcumin	Best pH 4.0–6.0	Not suitable for very acid drinks
Matcha/green tea extracts	Best pH 4–5	Brown at low pH
Collagen peptides	Best pH 3.6–4.0	Aggregates below 3.3
Botanical extracts	pH 3–5	Polyphenol precipitation possible

Sweeteners	Good stability	Stevia stable; sucralose stable at low pH

E3.2: Shelf-Life Impact Factors

Factor	Effect on Stability	Control Strategy
pH	Drives clarity, preservation	Maintain 3.5–3.8
Dissolved oxygen	Oxidation of vitamins/flavors	Deaeration + N_2 flush
Light exposure	Color & vitamin degradation	Amber packaging
Temperature spikes	Aroma & nutrient loss	Controlled logistics
Mineral interactions	Clarity & taste issues	Use organic salts
Osmolality	Mouthfeel, gastric comfort	Match use case

E3.3: Stability Diagnostic Matrix

Observed Issue	Likely Root Cause	Corrective Action
Haze formation	Polyphenol–protein complex	Add pectin; adjust pH to 3.8
Color shift to brown	Oxidation	Add ascorbate + chelator; reduce DO
Sediment at bottom	Undissolved salts/proteins	Increase mixing or change salt form
Metallic notes	KCl or Mg^{2+} concentration too high	Flavor masking; reduce dose
Vitamin loss	High heat or oxygen	Post-pasteurization dosing

E4: Rapid Formulation Decision Flowcharts

E4.1: pH Adjustment Decision Tree

1. Is product too acidic (<3.3)?
   → Raise with sodium or potassium citrate.

2. Is product not acidic enough (>3.9)?
   → Lower using citric or malic acid.

3. Is clarity affected?
   → Aim for 3.6–3.8.

4. Does taste feel sharp?
   → Use malate + citrate blend for smoother acidity.

E4.2: Osmolality Adjustment Decision Tree

1. Too high (>320 mOsm/kg)?
   → Reduce sugars, polyols, or mineral salts.

2. Too low (<200 mOsm/kg)?
   → Add sugar or organic acids for body.

3. Goal is fast hydration?
   → Target 150–240 mOsm/kg.

4. Goal is sustained recovery?
   → Target 280–330 mOsm/kg.

E4.3: Stability Optimization Quick Guide

Table E4.3: Stability Optimization Quick Guide

Desired Outcome	**Action**
Longer shelf life	Lower DO, lower pH, antioxidants
Better color retention	Amber packaging, nitrogen flush
Smoother flavor	Use citrate and malate blend
Less aftertaste	Add aroma top notes; use bitter blockers
Improved clarity	Control pH; use microfiltration

36.6 Appendix F: Supplier and Ingredient Directory

This directory provides a curated list of established global suppliers of key raw materials used in hydration, electrolyte, and functional nutrition beverages.
It is organized by ingredient category, with supplier strengths and typical applications to support procurement decisions, formulation consistency, and reliability.

This appendix serves as a starting point for ingredient sourcing; formulators should always confirm regional availability, regulatory compliance, and documentation (COA, MSDS, allergen statements, non-GMO declarations, stability data).

F1. Electrolytes and Mineral Salts

Magnesium, Potassium, Sodium, Calcium

Table F1.1: Electrolyte Suppliers

Supplier	Region	Key Strengths	Typical Ingredients
Jungbunzlauer	EU / Global	High-purity organic salts; consistent particle size	Citrates, lactates, gluconates (Na, K, Mg, Ca)
Cargill	USA / Global	Food-grade mineral salts; strong global logistics	Sodium citrate, potassium salts
Gadot Biochemical Industries	Israel / Global	Premium mineral chelates, clean-label focus	Calcium citrate, Mg citrate, K gluconate
Dr. Paul Lohmann	Germany	Pharma-grade mineral salts	Magnesium lactate, magnesium chloride, calcium salts
Corbion	Global	High-solubility organic acids & salts	Lactates; mineral lactates
Innophos	USA	High-quality phosphates	Sodium phosphate, potassium phosphate

Formulator's Note:

For clear beverages, prioritize gluconates and citrates; for stronger buffering systems, lactates may be preferred.

Table F1.2: Natural Flavors, Citrus Elements, and Botanical Extracts

Supplier	Region	Specialty	Notable Ingredients
Givaudan	Global	Premium, nuanced flavor systems	Citrus oils, tropical profiles, flavor modulators
Symrise	EU/Global	Aroma chemistry + naturals	Fruit flavors, cooling agents
Firmenich (DSM Firmenich)	Global	High-end beverages & wellness	Citrus, mint, botanicals
Takasago	Japan / Global	Asian flavor systems, green tea expertise	Matcha flavors, yuzu, lychee
Kerry	Global	Large-scale beverage flavoring	Juice-type flavors, sweet enhancers
Blue California	USA	Natural actives from fermentation	Stevia glycosides, botanical extracts
Botanica / Martin Bauer	EU/USA	Herbal extracts, tea extracts	Chamomile, ginger, hibiscus, adaptogens

Formulator's Note:

Botanical stability varies, request polyphenol content, pH compatibility, and heat tolerance data.

Table F1.3: Sweeteners and Sweetness Modulators

Supplier	Category	Key Ingredients
PureCircle (Ingredion)	Stevia experts	Reb M, Reb D, enzymatically modified stevia
Layn Natural Ingredients	Natural sweeteners	Monk fruit (mogrosides), stevia blends
Tate & Lyle	Global sweetener leader	Allulose, sucralose, stevia, sweetness enhancers
CJ CheilJedang	Fermentation-based actives	Allulose, rare sugars
ADM	Sweetness systems	Allulose, sucralose, stevia blends
Mafco Worldwide	Taste modulator	Licorice derivatives for bitterness masking

Formulator's Note:

For hydration beverages, stevia + monk fruit blends provide clean label sweetness with low osmolality impact.

F4: Functional Ingredients: Vitamins, BCAAs, Adaptogens, Nootropics

Table F4.1: Suppliers of Functional Ingredients

Supplier	Focus	Sample Ingredients
DSM (Firmenich DSM)	High-quality, stable vitamins	Vitamin C (specialty grades), B-complex, vitamin D
Lonza / Capsugel	Amino acids & actives	BCAAs, L-carnitine
Kyowa Hakko	Fermentation-pure actives	BCAAs, L-theanine, glutathione
Sabinsa	Botanical actives	Curcumin, ashwagandha, piperine
Nutraceuticals World-affiliated suppliers	Adaptogens/nootropics	Rhodiola extract, ginseng, bacopa
Glanbia Nutritionals	Protein and performance blends	Hydrolyzed proteins, amino blends

F5: Natural Colors (Anthocyanins, Carotenoids, Chlorophylls)

Table F5.1: Suppliers of Natural Colors

Supplier	Specialty	Notes
Chr. Hansen	Fermentation-derived natural colors	Stable anthocyanins
DDW (A Kerry Company)	Natural color experts	Carotenoids, betalains
GNT (EXBERRY)	Fruit & vegetable concentrates	Clean label, pH-dependent
Sensient	Wide portfolio	Blue spirulina, turmeric, anthocyanins

Formulator's Note:

pH directly shifts color tones for anthocyanins, request pH/color maps.

F6: Stabilizers, Texturizers, and Emulsion Tools

Table F6.1: Suppliers of Stabilizers, Texturizers, and Emulsion Tools

Supplier	Focus	Key Materials
CP Kelco	Polysaccharides	Pectin, gellan gum, xanthan gum
TIC Gums (IFF)	Texture & mouthfeel systems	Gum blends, stabilizer systems
ADM	Emulsifiers	Modified starches, natural emulsifiers
AEP Colloids	Specialty gums	Acacia gum, gelatin-based systems
Kerry	Beverage stabilizers	Cloud systems, texture enhancers

Formulator's Note:

For clear beverages, avoid viscous gums, use low-acyl gellan or light pectins.

F7: Concentrates, Juices, and Fermented Ingredients

Table F7.1: Suppliers of Concentrates, Juices, and Fermented Ingredients

Supplier	Strengths	Example Ingredients
Doehler	Beverage-grade concentrates	Citrus bases, fermented fruit essences
Kagome	Vegetable + fruit extracts	Carrot concentrates, tomato lycopene
Citrosuco	Citrus raw materials	Orange, lemon, lime
Archer Daniels Midland (ADM)	Large-scale juice systems	Clarified juices, essences

F8: Packaging Materials and Barrier Systems

Table F8.1: Suppliers of Packaging Materials and Barrier Systems

Supplier	Category	Notes
Amcor	PET, recycled PET	Lightweight, high-clarity bottles
Tetra Pak	Cartons	For aseptic systems
Aptar	Closures & dispensing	High-barrier closures
SIG	Aseptic carton systems	Good for shelf-stable beverages

Formulator's Note:

High-acid products benefit from PET with oxygen scavengers.

F9: Co-Manufacturers and Pilot-Scale Facilities

(General categories; specific companies vary by region.)

Table F9.1: General categories of Co-Manufacturers and Pilot-Scale Facilities

Region	Co-Manufacturing Strength
USA	RTD hydration, vitamin waters, high-speed aseptic
Europe	Clean-label natural beverages, organic-certified lines
Asia	Green tea beverages, electrolyte concentrates, fermentation ingredients
Middle East / GCC	Isotonic drinks, PET bottling, energy drinks
Latin America	Juice blends, tropical beverages, powders

F10: Documentation Checklist (For Every Supplier)

Before approving any supplier, obtain:

- Certificate of Analysis (COA)
- Safety Data Sheet (SDS)
- Allergen statement
- Non-GMO / GMO-free documentation
- Vegan / halal / kosher certifications (if applicable)
- Shelf-life and storage data
- Microbiological specifications
- Heavy metals report
- Stability data in acidic beverages (for vitamins, botanicals)
- Origin statement (country of manufacturing and processing)

www.ingramcontent.com/pod-product-compliance
Lightning Source LLC
LaVergne TN
LVHW082024200926
848481LV00010B/2918